Biotechnology Monographs

Volume 3

Editors

S. Aiba · L.T. Fan · A. Fiechter · J. Klein · K. Schügerl

L.T. Fan · M. M. Gharpuray · Y.-H. Lee

Cellulose Hydrolysis

With 65 Figures

Springer-Verlag
Berlin Heidelberg New York
London Paris Tokyo

Prof. Liang-tseng Fan
Department of Chemical Engineering
Durland Hall, Kansas State University
Manhatten, KS 66506, USA

Dr. Mahendra Moreshwar Gharpuray
Group Operations, Inc.
1110 Vermont Ave, N.W., Suite 500
Washington, D.C. 20005, USA

Prof. Yong-Hyun Lee
Department of Genetic Engineering
College of Natural Sciences,
Kyungpook National University
Taegu 635, Korea

ISBN-13: 978-3-642-72577-7   e-ISBN-13: 978-3-642-72575-3
DOI: 10.1007/978-3-642-72575-3

Library of Congress Cataloging in Publication Data.
Fan, L.T. (Liang-tseng), Cellulose hydrolysis. (Biotechnology monographs; v. 3) In-
cludes bibliographies and indexes. 1. Cellulose – Congresses. 2. Hydrolysis – Congresses.
I. Gharpuray, M.M. (Mahendra Moreshwar). II. Lee, Y.-H. (Yong-Hyun), III. Title.
IV. Series. QD320.F36 1987   547.7'82   87-20552

To
Dr. MARY MANDELS
for
Her Pioneering Research
in
Cellulose Hydrolysis

Foreword

Recent economic trends, especially the worldwide decline in oil prices, and an altered political climate in the United States have combined to bring about major reductions in research on renewable energy resources. Yet there is no escaping the "facts of life" with regard to these resources. The days of inexpensive fossil energy are clearly numbered, the credibility of nuclear energy has fallen to a new low, and fusion energy stands decades or more from practical realization. Sooner than we may wish we will have to turn to renewable raw materials – plant "biomass" and, especially, wood – as significant suppliers of energy for both industry and everyday needs. It is therefore especially important to have a single, comprehensive and current source of information on a key step in any process for the technological exploitation of woody materials, cellulose hydrolysis. Furthermore, it is essential that any such treatment be unbiased with respect to the two methods – chemical and biochemical – for the breakdown of cellulose to sugars.

Researchers on cellulose hydrolysis have frequently been chided by persons from industry, especially those individuals concerned with determining the economic feasibility of various technological alternatives. They tell us that schemes for the utilization of wood and other such resources fly in the face of economic realities. The proper response to such allegations was offered by one of this country's best known "wood" chemists at one of the symposia on "Biotechnology for Fuels and Chemicals" (sponsored by the Oak Ridge National Laboratory). In a free-form discussion on this subject, Professor Irving Goldstein (North Carolina State University) pointed out that it is not the function of the research community to follow the dictates of perceived economic feasibility in choosing subjects for research. Our function is to provide the knowledge base on which technologies may be fashioned. Then, when the times are ripe, the knowledge needed to take advantage of different options will be at hand. This viewpoint provides a cogent argument for the volume which follows.

August 1987

ELMER L. GADEN, Jr.
Wills Johnson Professor and Chairman
Department of Chemical Engineering
University of Virginia

Preface

This monograph covers numerous facets of cellulose hydrolysis. Presentation is expositional so that the book will be useful not only as a reference monograph but also as a textbook. The monograph is comprehensive in that both enzymatic hydrolysis and acid hydrolysis are treated, and their fundamental and applied aspects are discussed as well. Physical and chemical properties and characteristics of cellulose pertaining to these hydrolysis processes are also elucidated. In addition to qualitative elaboration of each subject, every effort is made to describe it quantitatively or mathematically.

Specifically, this monograph furnishes a compendium of the nature of cellulosic materials including their compositions and structures; it elaborates on the nature of lignocellulosic structural resistance, properties and mode of enzymatic action, different pretreatment methods, and a variety of kinetic models for enzymatic degradation; it focuses on the mechanism and kinetics of acid hydrolysis; and finally it describes commercial hydrolysis processes for a variety of cellulosic materials.

Acknowledgement

The authors' research, supported for several years by the U.S. Department of Energy and Kansas State University, has culminated in the present monograph. Besides the authors, Mr. David Beardmore and Mr. Bob Wisecup participated in the research; naturally, they have made significant contributions. While the manuscript was typed by several secretaries, the authors wish to single out Mrs. Janet Vinduska. The first author (LTF) also wishes to acknowledge the aid of his wife, Eva, in preparing the manuscript.

August 1987
L. T. Fan
M. M. Gharpuray
Y.-H. Lee

Table of Contents

1 Introduction

Cellulose is an abundantly available carbohydrate polymer in nature. This polymer is continually replenished by photosynthetic reduction of carbon dioxide catalyzed by sunlight. The estimated volume of existing cellulosic resources is 324 billion cubic meters. Furthermore, the annual net yield of photosynthesis is 1.8 trillion tons of biodegradable substances, about 40% of which is cellulose. The abundance coupled with the renewability renders cellulose to be the most promising feedstock for production of energy, food, and chemicals. In addition, the nature's balance and aesthetic value are not diminished by its use as feedstock [4]. Only a small fraction of cellulosic resources are currently utilized to manufacture products such as lumber, fuel, textiles, paper, plastics, films, foils, explosives, varnishes, thickeners, and glues. Moreover, easily harvestable trees, which are rich in cellulose, tend to be exploited for this purpose [2].

In addition to cellulose present in trees, vast amounts of cellulose exist in cellulosic wastes, such as municipal waste, agricultural wastes, and animal manure. Municipal wastes include paper and packaging materials; they have served as landfills or fuel sources to generate steam through combustion or through pyrolysis. Combustion and pyrolysis are attractive alternatives for utilizing cellulose; however, these processes can not handle excessively wet wastes such as municipal sewage or pulping wastes. Agricultural wastes generally arise from food production where food grains are harvested and associated plant materials are discarded. Food wastes, including coffee grounds, oat hulls, potato and fruit peelings, are also generated from food processing. Presently, animal manure is either returned to the land as fertilizer or digested by anaerobic fermentation to produce methane gas. The production of methane is hampered by drawbacks including the high retention time (10–30 days) and instability of the microbial system involved in the fermentation.

In order to utilize cellulose as a feedstock for production of energy, food and chemicals, its degradation to glucose is essential. Cellulose is resistant to degradation since a cellulose molecule serves largely as a structural molecule rather than as a molecule to store energy [4]. It is reported that theoretically about 77 kcal of incident radiational energy is stored in each gram of pure cellulose. Unfortunately, the majority of existing processes utilizing cellulose extract only a small fraction of this energy due to their high energy consumption.

Cellulose is a polymer composed of anhydroglucose monomer units. Its degradation to glucose is accomplished by addition of a water molecule for each glucose molecule produced. The potential of this reaction, termed hydrolysis, has been well stated by Hajny and Reese [5]. "Cellulose is the major constituent of all

1

vegetation, comprising from one-third to one-half of dry plant material. As such it is one of the world's most plentiful resources. ... Most vegetation is unused by man or animals and undergoes natural decay by microorganisms capable of producing cellulases. ... Among the many problems facing the world today, two of the most important are pollution and the threat of famine. ... With the tremendous world-wide activity in seeking foodstuffs from conventional and unconventional sources, it is surprising that cellulose and lignocellulose have received so little attention. ... (the) amount of scientific research devoted to finding practical means to produce a digestible nutrient from cellulose bearing materials is negligible. Now is a propitious time for attacking this problem. Success can not only alleviate hunger, but by increased utilization of residues, it can assist in pollution abatement."

The hydrolysis of cellulose to glucose can be catalyzed by an enzyme or an acid. Enzymatic hydrolysis has been investigated extensively in the last decade or so. At present, no commercial plant appears to exist for enzymatic hydrolysis. A pilot plant to enzymatically consume dried bagasse and rice straw has been built in Japan [1]. Acid hydrolysis has been industrially practiced in the U.S. since 1913 when a plant was constructed in South Carolina to hydrolyze waste from pine mills. The plant eventually became uneconomical due to a decrease in prices of other sugar sources. In Germany, acid hydrolysis was performed using dilute sulfuric acid and dilute hydrochloric acid during World War II [7].

Enzymatic hydrolysis has several advantages over acid hydrolysis. The enzymes hydrolyzing cellulose are very specific and do not produce undesirable by-products; at the end of hydrolysis, neutralization of the product is not required; capital and operating costs are substantially lower since enzymatic hydrolysis takes place at mild temperatures, and expensive corrosion-proof equipment is not required. Furthermore, recent advances in genetic engineering have yielded highly effective strains of cellulose enzymes, thereby bringing the enzymatic hydrolysis process nearer to commercially viability. The advantages of acid hydrolysis over enzymatic hydrolysis include short reaction time and the ability to utilize cellulose without expensive pretreatment [3].

The main product of hydrolysis, namely glucose, commands a wide range of usage.
(a) Its refined form can be consumed directly as a food source in the food industry.
(b) Glucose subject to isomerization becomes a sweetner.
(c) It can be a substrate for alcohol fermentation.
(d) It can be a substrate for single cell protein production.
(e) It can be a substrate for biopolymer production [6].
(f) It can be a substrate for other fermentations in the production of chemicals and drugs, such as Penicillin, gluconic acid, L-amino acids, citric acid, pectin, aetone, and butanol.

2

References

1. Anonymous (1984) Chemical Week 135 (2):48
2. Casey JP (1980) Pulp and paper, 2nd edn, vol 1. Interscience, New York
3. Converse AO, Grethlein HE (1984) Chem Eng Ed Fall, p 186
4. Ghose TK (1977) In: Ghose TK et al. (eds) Adv Biochem Eng, vol 6. Springer, Berlin Heidelberg New York, p 39
5. Hajny GJ, Reese ET (1969) Preface. Adv Chem Ser 95
6. Shipman RH, Fan LT (1978) Proc Biochem 13 (3):19
7. Wenzl HFJ (1970) The chemical technology of wood. Academic Press, New York

3

2 Nature of Cellulosic Material

The hydroysis of native lignocellulosics, especially that catalyzed by enzyme, is a slow process. The heterogeneous degradation of lignocellulosics is governed primarily by their structural features since (1) cellulose present in biomass possesses a highly resistant crystalline structure, (2) lignin surrounding the cellulose forms a physical barrier, and (3) the reactive sites available are limited. The cellulose present in lignocellulosics is composed of crystalline and amorphous components [7]. The amorphous component is usually more reactive than the crystalline component, and thus any means that will increase the amorphous content will enhance the hydrolysis rate [12, 16, 17, 50]. The presence of lignin forms a physical barrier for attack by either enzyme or acid molecules; therefore, treatments causing disruption of the lignin seal will increase the accessibility of cellulose to enzyme or acid molecules and eventually its hydrolysis rate. The limitation of available reactive sites stems from the fact that the average size of the capillaries in biomass is too small to allow the entry of reactive molecules, especially the large enzyme molecules.

This chapter furnishes a compendium of the nature of cellulosic materials including their compositions and structures.

2.1 Components of Cellulosic Materials

Cellulosic materials are composed of three major components, extraneous substances, polysaccharides, and lignin, as illustrated in Fig. 2.1 [28]. These components are delineated below.

2.1.1 Extraneous Materials

The extraneous component refers to all the non-cell wall materials. This component consists of an astonishingly wide variety of chemicals. Based on their solubilities in water and neutral organic solvents, these chemicals can be classified as extractives or non-extractives [28].

The extractives can be crudely divided into three groups, namely, terpenes, resins, and phenols [44]. The terpenes are regarded as isoprene polymers and are a source of turpentine in industrial processes. Related to terpenes are terpene alcohols

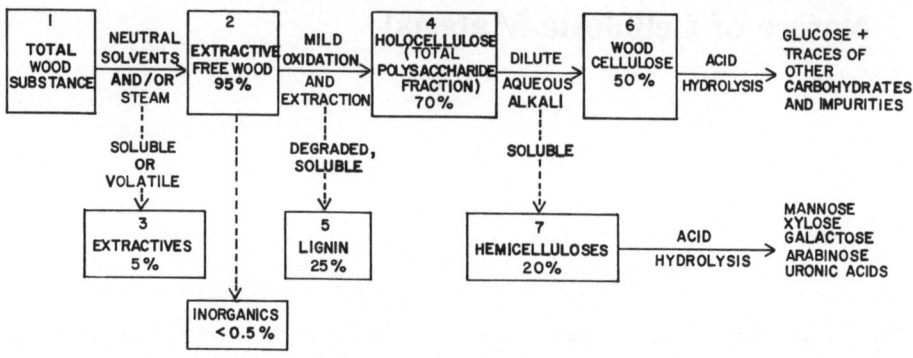

Fig. 2.1. Classification of the major components of wood [28]

and ketones. The resins include a wide variety of non-volatile compounds, including fats, fatty acids, alcohols, resin acids, phytosterols, and less known neutral compounds in small amounts. The phenols consist of numerous compounds which have not been explored adequately; the most important among them are tannins, heartwood phenols, and related substances. Other extractives include low molecular weight carbohydrates, alkaloids, and soluble lignin.

The non-extractives mainly consist of inorganics mostly present in ash minerals. The dominating components are alkali and alkali earth carbonates and oxalates. Silica deposited as crystals is especially abundant in straws; furthermore, small amounts of non-cell wall substances, such as starch, pectin, and protein, are not extractable [28].

In spite of the fact that they exist in small quantities, extraneous compounds play a very significant role in that they render cellulose not only resistant to decay and insect attack but also inhibitive to pulping and bleaching.

2.1.2 Polysaccharides

The polysaccharide component comprises high molecular weight carbohydrates, namely, cellulose and hemicellulose, which amount to 60 to 80 % of the total wood. Cellulose is the major component of cell walls of wood fiber, and it is a linear polymer of D-glucose with a high molecular weight of approximately half a million. Individual glucose molecules are linked together by β-1,4 linkages to form a highly crystalline material that is resistant to enzymatic hydrolysis [16]. The number of chain units, the so-called degree of polymerization or DP in short, varies for different cellulosic materials (Table 2.1). The glucosidic bonds in chain molecules, along with the hydroxyl groups, mainly determine its chemical properties [16]. The innumerable hydrogen bonds, holding the chains together, are not broken by water; cellulose is completely insoluble in water. However, strong acids, strong alkalis, concentrated salt solutions, and various complexing reagents can swell or disperse and even dissolve the cellulose [23].

Table 2.1. Molecular weight and degree of polymerization of cellulosic materials

Cellulosic material	Molecular weight	Degree of polymerization
Before hydrolysis [43]		
Native cellulose	600,000–1,500,000	3,500–10,000
Chemical cottons	80,000– 500,000	500– 3,000
Wood pulps	80,000– 340,000	500– 2,100
Rayon filament	57,000– 73,000	
After hydrolysis with 2.5 N HCl at 105 °C for 15 min [6]		
Natural fibers:		
Ramie, hemp		350–300
Cotton, purified		250–200
Unbleached sulphite wood-pulp		400–250
Bleached sulphite pulp		280–200
Mercerized cellulose (18% NaOH at 20 °C, 2 h)		90– 70
Vibratory milled wood cellulose		100– 80
Regenerated fibers:		
Fortisan, fiber G		60– 40
Textile yarns		50– 30
Tire yarns		30– 15

Hemicellulose is composed of shorter chain polysaccharides, and it is the principal non-cellulosic fraction of polysaccharides. The role of this component is to provide a linkage between lignin and cellulose. In its natural state, it exists in an amorphous form and can be divided into three groups, namely, xylans, mannans and galactans; these groups can exist separately as single components or collectively [4, 9, 29, 37]. The xylans are present as arabinoxylans, glucuronoxylans, or arabinoglucuronoxylans; the mannans are present in wood as glucomannans and galactomannans; and the galactans are relatively rare but are often found in the form of arabinogalactans [29, 41].

Xylan is a polymer of 4-O-methylglucuronoxylan and 4-O-methyl-glucuronoarabinoxylan linked by a β-D-(1 → 4)-bond, similar to the linkages of the glucose units in cellulose. In xylan, however, the hydroxyl groups are substituted with 4-O-methylglucuronic acid, arabinose and acetyl groups [35].

The mannan groups in the glucomannans from deciduous wood and coniferous wood are different; the former is characterized by the absence of galactose units. Thus, it is assumed that the glucomannans have a linear structure so that there is no branching at either of the two carbon positions, namely, (C-2) or (C-3). The presence of short side-chains of galactose units in the glucomannans from coniferous woods may be considered to be certain. Linnell and Swenson [32, 33] also have proposed that the configuration of the glucomannan from black spruce has a linear polysaccharide structure and that the glucomannan and lignin occur in the fiber in the form of a cross-linked matrix with a considerable number of lignin-carbohydrate bonds.

A possible structure of the arabinogalactan from tamarack had been proposed by Adams [1], based on the evidence of methylation and periodate oxidation. Later,

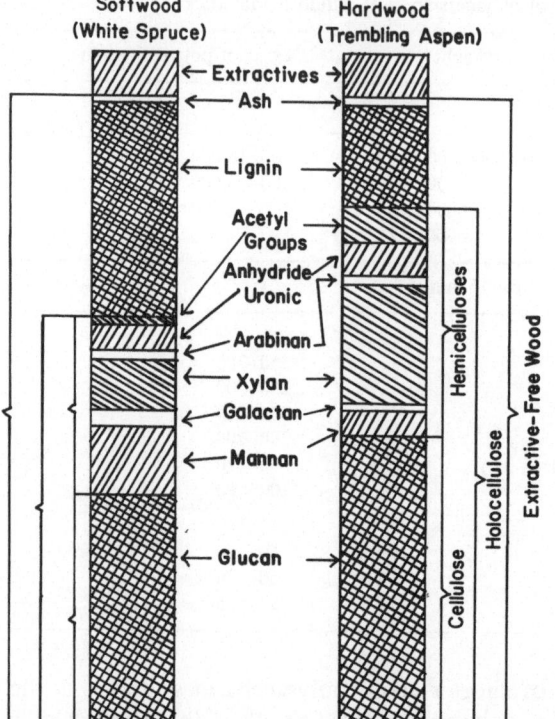

Fig. 2.2. Comparison of the compositions of hardwood and softwood [8]

Haq and Adams [25] have proposed a further modification to its structure. A possible structure for repeating units in galactans from other larch species has been proposed by White [54] and Aspinall et al. [5]. According to White [54], this structure is characterized by a main chain of 1, 6 linked galactose units substituted in position 3 by a secondary chain of three 1 → 3 linked D-galactose residues. According to Aspinall et al. [5], the structure consists of a 1 → 3 linked D-galactose main core with two 1 → 6 linked D-galactose units attached to it through position 6. Hardwood hemicelluloses are rich in xylan polymers with small amounts of mannan, whereas softwood hemicelluloses are rich in mannan polymers and contain significant quantities of xylans, as shown in Fig. 2.2 [8].

2.1.3 Lignin

Lignin is probably the most complex and least well characterized molecular group among wood components [48, 52]. It is essentially a three dimensional phenylpropane polymer with phenylpropane units held together by ether and carbon-carbon bonds [20]. The amount of lignin constitutes 20–35 % of the wood structure. Lignin possesses a high molecular weight and is amorphous in nature. In wood, the lignin network is concentrated between the outer layers of fibers as illustrated in

8

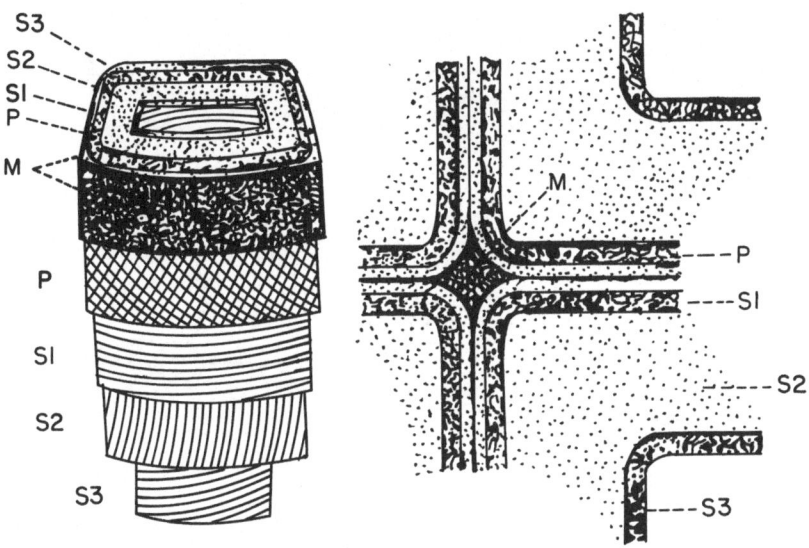

Fig. 2.3. Diagrammatic sketch of the intricate structure of wood cell wall [12]

Fig. 2.3. The lignin gives structural rigidity by stiffening and holding the fibers of polysaccharides together [13].

The lignin from grasses, softwoods, and hardwoods differs somewhat in composition, mainly in methoxyl substitution, and the degree of linkage between phenyl groups. Nevertheless, their common structural features predominate [27]. Figure 2.4 shows the schematic structure for spruce lignin [3].

The arylglycerol-β-aryl ether structure (Units 1–2, 2–6, 6–7, 7–8, etc. in Fig. 2.4) has been found to be the most abundant interphenylpropane linkage in lignin [2, 15, 26, 30, 45]. It plays a dominant role as a connecting link between the phenylpropane structural units. The phenylcoumaran structures (Units 4–5 and 15–16 in Fig. 2.4) have been identified by Sakakibara [45] and Freudenberg [21], respectively. The diarylpropane structures (Units 3–4 in Fig. 2.4) have been isolated as hydrolysis products of coniferous [47] and hardwood lignin [34, 38].

Ericksson et al. [15] and Larsson-Miksche [30] have estimated the amount of biphenyl structure (Units 11–12 and 16–18 in Fig. 2.4) in lignin as being approximately 0.095–0.11 per C_6–C_3 unit. Diphenyl ether compounds, containing the 4-O-5 linkage (Units 8–9 and 13–16 in Fig. 2.4), have been isolated [22, 39, 46]. The amount of this structure has been estimated to be approximately 0.035–0.04 per C_6–C_3 unit [15].

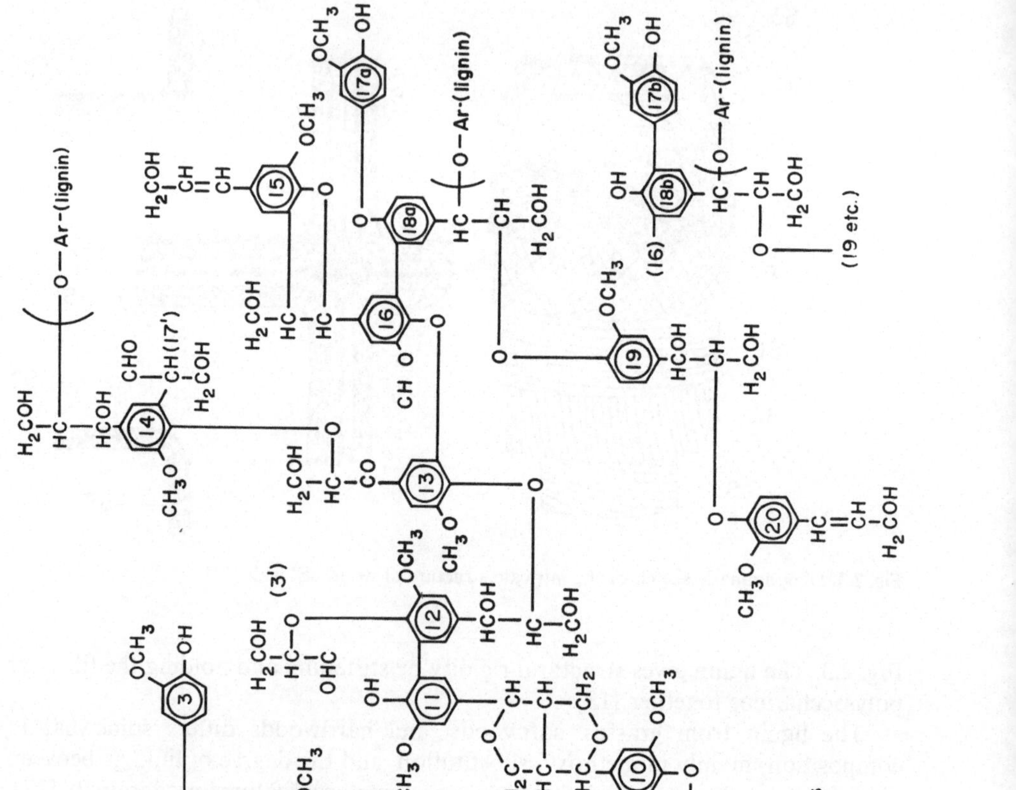

Fig. 2.4. Schematic molecular structure of spruce lignin [3]

10

2.2 Composition of Wood and Straw

Table 2.2 presents the compositions of several hardwood and softwood species [10]. As depicted in Fig. 2.2 [8], the compositions of hardwoods and softwoods are significantly different. It can be seen from the figure that: (1) the lignin content of softwoods is generally higher than that of hardwoods, (2) the hemicellulose content of hardwoods is similar to that of softwoods, and (3) the cellulose content of hardwoods is generally higher than that of softwoods. It must be kept in mind, however, that wide variations in chemical composition occur not only between different species but also within a single species.

Table 2.3 presents the compositions of various types of straws [10]. Straw species are more uniform in composition than various wood species. Generally, straw has a lower cellulose content than wood, but in spite of this, it has a total carbohydrate fraction (holocellulose) approximately equal to that of wood. This is due to its high hemicellulose and low lignin contents compared to wood. The ash content is greater in straw than in wood.

2.3 Structure of Plant Cell Walls

The intricate structure of a wood cell wall is schematically depicted in Fig. 2.3 [12]. It shows the primary wall (P), the thin outer layer of the secondary wall (S1), the substantial middle layer (S2), and the very thin inner layer (S3) sometimes called the tertiary wall [12–14].

Within each layer of the secondary wall (S), the cellulose and other cell wall constituents are aggregated into long slender bundles called microfibrils. The microfibrils are distinct entities in that few cellulose molecules, if any, cross over from one microfibril to another. In the (S1) layer, the microfibrillar groups are in helixes alternately crossed. In the middle layer (S2), the microfibrillar groups are oriented in bands (lamellae) nearly parallel to the cell axis; and in the inner layer (S3) the direction is nearly perpendicular to that in (S2). The primary wall (P) has an irregular helical arrangement around the cell axis. Surrounding the fiber is the heavily lignified and stiff middle lamellae (M), shared by adjacent fibers [12, 14].

Figure 2.5 shows the distribution of the chemical constituents in a typical cell wall [24]. The middle lamellae is heavy in lignin and is about 1–2 μ thick; it is amorphous and generally porous. The primary wall (P) is usually very thin and remains so throughout the growth of the plant, whereas the secondary wall (S) thickens during cell growth and contains most of the cellulose.

In a general construction scheme around a hollow lumen, fibrillar elements consisting of cellulose are wound in spirals to give the fiber its tenacity and flexibility. Furthermore, this "cable" is made waterproof by lignin or waxy compounds, which offer chemical resistance. Hemicellulose provides an intimate interlacing and even bonding between the lignin and cellulose. The compounding is perfect; linear but hydrophilic cellulose chains contribute tensile strength, while

Table 2.2. Chemical composition of some Eastern and Lake States' species[a] [10]

Species	Chemical composition of wood (moisture free basis), %						
	Holocellulose		Cross and Bevan cellulose		Lignin	Total pentosans	Ash
	Total	Alpha	Total	Alpha			
Hardwoods							
Quaking aspen (*Populus tremuloides*)	82	51	64	48	17	23	0.3
American beech (*Fagus grandifolia*)	78	–	61	47	23	20	0.2
Paper birch (*Betula papyrifera*)	–	–	60	41	25	26	–
Yellow birch (*B. lutea*)	–	–	61	–	–	25	0.5
Eastern cottonwood (*Populus deltoides*)	–	–	63	46	24	19	–
Sugar maple (*Acer saccharophorum*)	76	–	57	42	23	21	0.2
Silver maple (*A. saccharinum*)	83	–	61	47	21	18	0.2
Yellow poplar (*Liriodendrum tulipifera*)	–	–	62	45	20	19	–
Black cherry (*Prunus serotina*)	85	–	60	45	21	20	0.1
White oak (*Quercus alba*)	–	–	51	–	32	22	0.4
Softwoods							
Balsam fir (*Abies balsamea*)	70	44	58	42	29	11	0.5
Eastern hemlock (*Tsuga canadensis*)	68	48	56	43	32	10	0.4
Jack pine (*Pinus banksiana*)	72	49	58	41	27	13	–
Eastern white pine (*P. strobus*)	–	–	60	44	28	11	–
Red pine (*P. resinosa*)	–	–	54	–	24	11	–
Black spruce (*Picea mariana*)	–	–	61	44	27	11	0.3
Red spruce (*P. rubra*)	73	48	60	43	27	12	0.2
White spruce (*P. glauca*)	73	49	61	44	27	11	0.3
Tamarack (*Larix laricina*)	–	–	–	–	–	–	–

[a] Typical values obtained at Forest Products Laboratory and thus wide variations may occur within single species.

Table 2.3. Composition of various types of American straws[a] [10]

Source / Component	Rice	Barley	Wheat	Rye	Oat	Flax shives	Soybean stalks
Moisture	8.0	8.4	6.6	7.4	7.0	8.1	8.3
Ash	16.1	6.4	6.6	4.3	7.2	3.5	2.3
Extractives							
Alcohol-benzene	4.6	4.7	3.7	3.2	4.4	4.1	3.9
Cold water	10.6	16.0	5.8	8.4	13.2	9.7	7.3
Hot water	13.3	16.1	7.4	9.4	15.3	11.4	8.8
1% NaOH	49.1	47.0	41.0	37.4	41.8	32.0	32.0
Nitrogen	0.58	1.10	0.38	0.72	0.46	0.69	0.66
Lignin	11.9	14.5	16.7	19.0	17.5	22.3	19.8
Pentosans	24.5	24.7	28.2	30.5	27.1	23.6	24.8
Cross and Bevan cellulose							
Crude	49.8	48.2	54.4	54.9	53.6	48.4	50.1
Ash free	48.6	47.4	53.6	54.3	53.4	46.2	48.2
Pentosans	27.8	30.0	26.8	29.5	28.4	21.6	22.1
Alpha cellulose	36.2	33.8	39.9	37.6	39.4	34.9	34.5

[a] All values except moisture given on % basis of moisture free material.

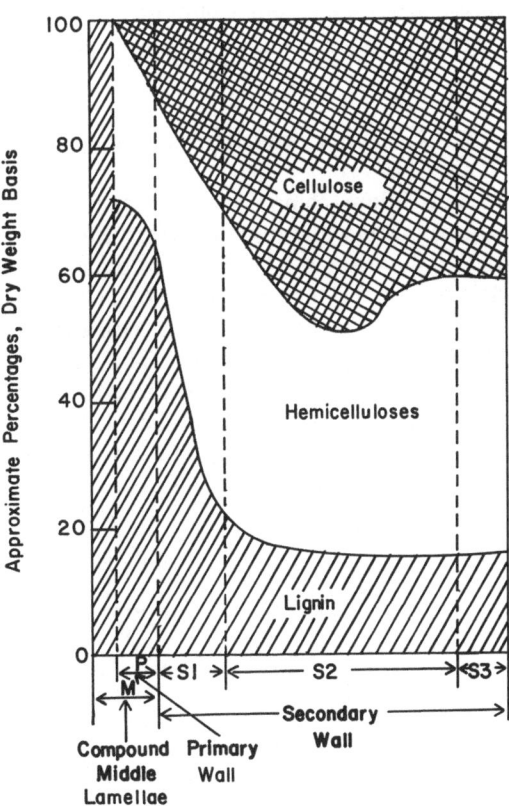

Fig. 2.5. Chemical constituents and their distribution in the wood cell wall [24]

13

hydrophobic amorphous lignin or similar compounds bring about chemical resistance, and in particular, protection against water [48].

The cell wall structure of cereal straw has not been extensively studied as compared to that of wood [10]. Compared to wood, straw is a much more heterogeneous raw material. Straw fibers, which are principally derived from cells and internodes, are fairly long and slender with sharply pointed ends. In addition to these fibers, however, straw also contains short nonfibrous cells consisting of epidermal cells, platelets, serrated cells, and spirals which are derived from the pitch, nodes, chaff, and rachises (head) [10]. Whereas 96% of the cells in wood may be considered as fibers, only about 35 to 39% of the cells in straw are fibers.

2.4 Structure of Cellulose

Cellulose, a high molecular weight linear polymer, is composed of D-glucose building blocks, joined by β-1,4-glucosidic bonds. In native cellulose, up to 10,000 β-anhydroglucose residues are linked to form a long chain molecule. This means that molecular weight of native cellulose is above 1.5 million. As the length of the anhydroglucose unit is 0.515 nm (= 5.15 Å), the total length of a native cellulose molecule is about 5 μm. Cellulose contained in pulp and filter paper usually has a degree of polymerization ranging from 500 to 2,100.

2.4.1 Chemical and Molecular Structure of Cellulose

In a cellulose chain molecule, each anhydroglucose unit assumes the chair configuration with the hydroxyl groups in the equatorial and the hydrogen atoms in the axial positions. The conformational formula (chair form) of cellulose (poly-1, 4-D-glucosan) is shown in Fig. 2.6; it can be seen that alternate chain unit is rotated 180° around the main axis. This results in an unstrained linear configuration, with minimum steric hindrance [49]. The glucosidic linkage acts as a functional group, and this, along with the hydroxyl groups, mainly determines the chemical properties of cellulose. All significant chemical reactions occur at these locations. It has been reported recently that the hydroxyl group in the 3-position is bound by an intramolecular hydrogen bond to the ring oxygen atom of the next chain unit [49].

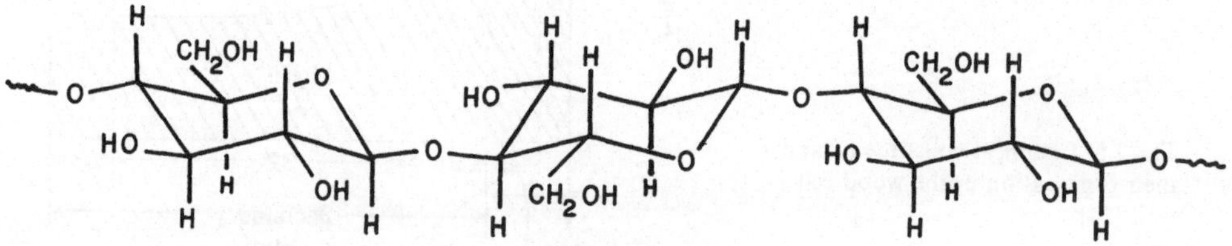

Cellulose

Fig. 2.6. The conformational form of cellulose [16]

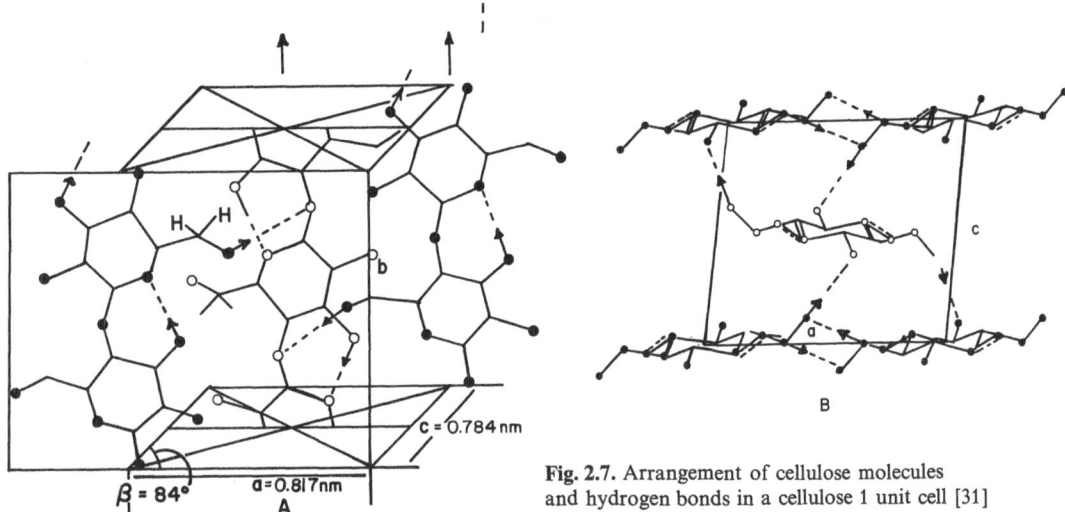

$\beta = 84°$ a=0.8l7nm A c = 0.784 nm

Fig. 2.7. Arrangement of cellulose molecules and hydrogen bonds in a cellulose 1 unit cell [31]

Cellulose molecules form a fibril, a threadlike long bundle of molecules, stabilized laterally by hydrogen bonding between hydroxyl groups of adjacent molecules. The molecular arrangement of this fibrillar bundle is sufficiently regular that cellulose exhibits a crystalline X-ray diffraction pattern. A schematic model of the native cellulose lattice has been proposed by Meyer, Mark, and Misch in 1929 [36]. Figure 2.7 gives a spatial representation of the unit cell or crystal of cellulose developed by Liang and Marchessault [31]. In this model, each unit cell contains four glucose residues. These include two in the center of the figure and one-fourth of each of the eight which are placed at the corners of a monoclinic cell. The corner residues are shared by each of the four unit cells meeting at the corners. The cellulose chain in the center of the unit cell runs in a direction opposite to that of the edges [49].

The dimensions of a unit cell and the strength of different bonds are listed in Table 2.4. The length of the unit cell, which is 10.3 Å, is the length of one of the repeating anhydrocellobiose units. Figure 2.7 also shows a projection of the unit cell on the ac-plane perpendicular to the direction of the b-axis. Notice that the shortest distance between atoms of neighboring chains of native cellulose is no more than 0.25 nm in the direction of the a-axis, which makes possible the formation of hydrogen bonds between adjacent chains. In the direction of the c-axis, the distance

Table 2.4. Stability of bonds in native cellulose [49]

Dimensions	Length nm	Stability kcal/mole	Nature of bond
a	0.817	15	Hydrogen
b	1.031	50	Covalent
c	0.784	8	van der Waals

15

is much greater and molecular chains are attached to each other by van der Waal's forces only [49].

Ward [51] has pointed out that the consequence of the high degree of order in native cellulose is that neither a water molecule nor an enzyme molecule can enter the structure. Therefore, the native cellulose is inert in the digestive tract. However, the structure of acid or alkali swollen cellulose is open and is split relatively easily by enzyme molecules.

2.4.2 Structure and Morphology of Cellulose Fibers

The chemical nature of cellulose, its physical and mechanical properties, and its fibrillary structure are deducible from its molecular structure. Like all hydrophilic linear polymers, individual cellulose molecules are linked together to form an elementary fibril or protofibril, about 40 Å wide, 30 Å thick, and 100 Å long, in which the polymer chains are oriented in a parallel alignment and firmly bound together by numerous strong hydrogen bonds. The elementary fibril is the smallest structural unit of microfibrils and fibers, and a number of the elementary fibrils are aggregated into a long slender bundle called a microfibril. In electron micrographs, the native cellulose microfibrils are usually seen as bundles in lamellae which contain an extremely large number of fibrillar units [42].

A schematic representation of the cross-section of a small lamellae of microfibrils, proposed by Rånby [42], is shown in Fig. 2.8. The crystalline region in which the linear molecules of cellulose are bonded laterally by hydrogen bonds is characterized by the cellulose lattice which extends over the entire cross-section of the microfibrils. This crystalline region is bounded by a layer of cellulose molecules that exhibit various degrees of parallelism. The less ordered region is called the paracrystalline or amorphous region. The specific dimensions of microfibrils and spatial relationship between the crystalline and paracrystalline regions are controversial subjects. The disordered region allows disintegration of the cellulose by hydrolysis into rod-like particles with aqueous, nonswelling, strong acid. Disordered areas of the microfibrils may be native or formed by mechanical forces, giving deformation beyond the limit of elastic recovery of the microfibrils. The

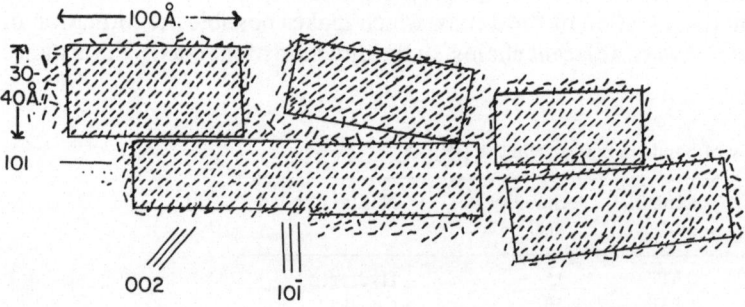

Fig. 2.8. Schematic representation of a lamellae of cellulose microfibrils from the secondary wall of a plant [42]

16

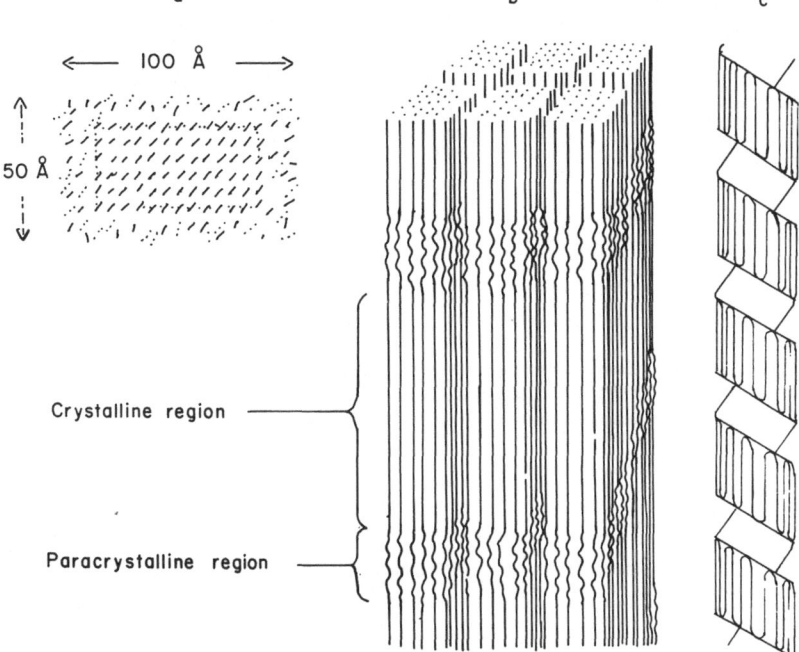

a   b   c

← 100 Å →

50 Å

Crystalline region

Paracrystalline region

Fig. 2.9. Recent concepts of the structure of cellulose microfibrils [12]

length of a resultant particle after acid hydrolysis (miscelles or microcrystals), which corresponds to the leveling-off-degree of polymerization of the hydrocellulose, varies with the extent of the pretreatment of native cellulose [42].

Cowling [12] reviewed recent models of microfibril structure, shown in Fig. 2.9, from which we can make the following three observations: (a) the microfibril is about 50×100 Å in cross-section and consists of a crystalline core of highly ordered cellulose molecules surrounded by a paracrystalline sheath. In cotton, this sheath contains mainly cellulose molecules but, in wood, it also contains hemicellulose and lignin molecules, (b) the cellulose molecules are less well ordered at certain points along the length of a microfibril, and (c) the cellulose molecules may exist in a folded chain lattice formed as a ribbon wound in a tight helix.

The fine structure of elementary fibrils (protofibrils) has been the subject of research for many years. Recently models of the various molecular arrangements of cellulose have been summarized by Chang [11]. These models are shown in Fig. 2.10. The term "LODP" in this figure stands for leveling-off-degree of polymerization. Two different types of models have been proposed from the molecular arrangement of elementary fibrils. On the basis of the observation of chain folding in a linear synthetic polymer, models of cellulose fibrils involving folded chains (models c, d, and e) have been suggested repeatedly. On the other hand, models based on extended cellulose molecules without folding (models a and b) are also reported in the literature. Chang [11] has concluded that model c is supported by most of the known facts about cellulose.

17

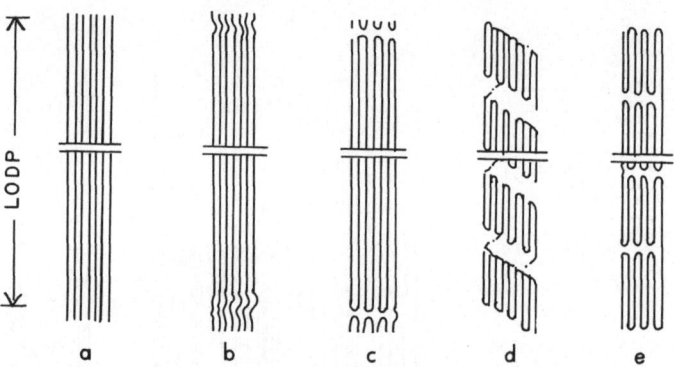

Fig. 2.10a–e. Models of molecular arrangements in the protofibril: **a** Full extension and complete crystalline model, **b** Fibrillized fringe miscellar model, **c** Fibrillized models of Ellefson, **d** Manley's planar zig-zag model, **e** Chain folding with fold length much smaller than LODP [11]

Model c (Fig. 2.10) has been incorporated into a high level structure, originally postulated by Fengel [18, 19], which is shown in Fig. 2.11. The figure shows the ultrastructural organization of the cell wall components. The elementary fibrils are cemented together by polyoses such as hemicellulose to form a microfibril. This microfibril is surrounded by a lignin and polyose layer which protects the microfibril from hydrolytic degradation. Pretreatment of cellulosics, aimed at degrading the protective layer, is essential, especially for a rapid enzymatic hydrolysis.

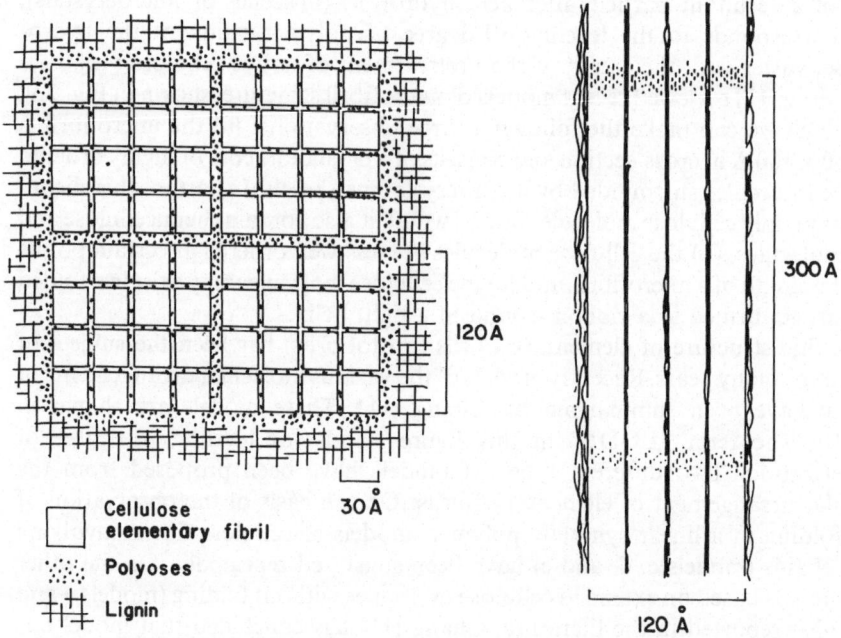

Fig. 2.11. Model of the ultrastructural organization of the cell wall components of wood [18]

18

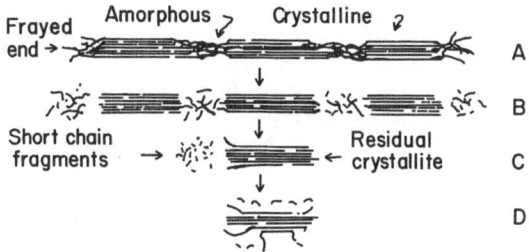

Fig. 2.12A–D. Schematic summary of initial stages of dissimilation of a partially crystalline cellulose fibril: **A** Original cellulose fibril, **B** Initial attack on the amorphous region, **C** Residual crystallite and digestion of short chain fragments, **D** Attack on residual crystallite [40]

The molecular structure of cellulose and the structures of an elementary fibril and a microfibril are important features that govern the hydrolytic degradation of cellulose. As mentioned previously, a cellulose fibril contains crystalline and amorphous regions. The proportion of crystalline or amorphous material has been estimated to range from 50 to 90 %. Generally, about 70 % crystallinity exists in native cellulose. Nisizawa [40] has depicted the different regions in a cellulose fibril at progressive stages of enzymatic degradation as shown in Fig. 2.12. In a partially crystalline cellulose fibril, the amorphous regions in the fringe miscelles are first attacked, leading to an enrichment of crystalline regions. The crystalline regions, in turn, are gradually solubilized after loosening their peripheral parts. Thus, the degradation of the substrate proceeds with mutual repetition of shortening of the chain length and subsequent loosening of the residual substrate. However, this scheme is highly idealized, considering the structural complexity of natural cellulose.

References

1. Adams GA (1960) Can J Chem 38:280
2. Adler E (1977) Wood Sci Technol 11:169
3. Adler E et al. (1969) Abst Intern Wood Chemistry Symp, Seattle, Washington, 1
4. Applegarth DA, Dutton GS (1965) TAPPI 48:204
5. Aspinall GO et al. (1958) J Chem Soc
6. Battista OA et al. (1956) Ind Eng Chem 48:33
7. Bikales NM, Segal L (1971) In: Cellulose and cellulose derivatives, parts 4 and 5. Wiley, New York
8. Browning BL (1970) In: Britt KW (ed) Handbook of pulp and paper technology, 2nd edn. Van Nostrand Reinhold, New York, p 7
9. Burman A (1953) J Text Res 23:888
10. Casey JP (1960) Pulp and paper, 2nd edn, vol 1. Interscience, New York
11. Chang W (1971) J Poly Sci, Part C, No 36, p 343
12. Cowling EB (1975) In: Wilke CR (ed) Biotechnol Bioeng Symp No 5, Interscience, New York, p 163
13. Cowling EB, Kirk TK (1976) In: Gaden EL, Jr et al. (eds) Biotechnol Bioeng Symp No 6, Interscience, New York, p 163
14. Cysewski GR, Wilke CR (1976) Biotechnol Bioeng 18:1297
15. Ericksson M et al. (1973) Acta Chem Scand 27:903
16. Fan LT et al. (1980) In: Fiechter A (ed) Adv Biochem Eng, vol 14. Springer, Berlin Heidelberg New York, p 101

17. Fan LT et al. (1980) Biotechnol Bioeng 22:177
18. Fengel D (1971) J Poly Sci, Part C, No 36, p 383
19. Fengel D (1970) TAPPI 53:497
20. Forss K, Fremer KE (1975) In: Bailey M et al. (eds) Symp on enzymatic hydrolysis of cellulose, SITRA, Helsinki, p 41
21. Freudenberg K (1968) Presented at constitution and biosynthesis of Lignin. Springer, Berlin Heidelberg New York, p 82
22. Freudenberg K, Chen CL (1960) Chem Ber 93:2533
23. Goldstein IS (1981) Organic chemicals from biomass, CRC Press, Florida
24. Hale JD (1969) In: MacDonald RG (ed) The pulping of wood, 2nd edn. vol 1, McGraw-Hill, New York, p 14
25. Haq S, Adams GA (1961) Can J Chem 39:1563
26. Higuchi T et al. (1973) Wood Research (Kyoto) 54:9
27. Irving SG (1976) In: Gaden EL Jr (ed) Presented at enzymatic conversion of cellulose materials: technology and applications. New York, p 293
28. Janes RL (1969) In: MacDonald RG (ed) The pulping of wood, 2nd edn, vol 1. McGraw-Hill, New York, p 34
29. Jayme G, Tio PK (1968) Das Papier 32 (6):322
30. Larsson LS, Miksche GE (1971) Acta Chem Scand 25:647
31. Liang CY, Marchessault RH (1959) J Poly Sci 37:385
32. Linnell WS, Swenson HA (1966) TAPPI 49 (10):444
33. Linnell WS, Swenson HA (1966) TAPPI 49 (11):491
34. Lundquist K, Miksche GE (1965) Tetrahedron Lettersxx:2131
35. Marchessault RH (1964) In: Proceedings Symp Grenoble, Imprimeries Reunies De Chamberg, p 287
36. Meyer KH et al. (1929) Z Physik Chem B2:115
37. Mitchell RL et al. (1956) TAPPI 39:571
38. Nimz H (1965) Chem Ber 98:3160
39. Nimz H (1971) ibid. 104:2539
40. Nisizawa K (1973) J Ferment Technol 51:267
41. Pecency R (1967) Svensk Paperstidn 70 (21):719
42. Rånby B (1969) Adv Chem Ser 95:134
43. Reese ET et al. (1972) Ghose TK et al. (eds) Presented at Adv in Biochem Eng, vol 2. Springer, Berlin Heidelberg New York, p 181
44. Rydholm SA (1965) In: Pulping processes. Interscience, New York, p 218
45. Sakakibara A (1977) In: Loewus FA et al. (eds) Presented at Recent Adv in Photochemistry. Plenum Press, New York, p 117
46. Schoenemann K (1954) In: FAO Technical panel on wood chemistry, Stockholm (1953); FAO Report 54/2/767
47. Schulz GV (1942) Z Physik Chem B52:50
48. Schurz J (1978) In: Ghose TK (ed) Bioconversion of cellulosic substances into energy chemicals and microbial protein symp proc. IIT Delhi, New Delhi, p 37
49. Sihtola H, Neimo L (1975) In: Bailey M et al. (eds) Symp on Enz Hyd of Cellulose. SITRA, Helsinki, p 9
50. Tsao GT et al. (1978) In: Perlman D (ed) Annual reports on fermentation processes, vol 2. Academic Press, New York, p 1
51. Ward KJr (1969) Schultz HW et al. (eds) Symp on Foods: carbohydrates and their roles. Conn.: Westport, AVI Pub, p 55
52. Wardrop AB (1971) In: Sarkanen KV, Ludwig CH (eds) Lignins. Wiley, New York, p 19
53. Wenzl HFJ (1970) The chemical technology of wood. Academic Press, New York
54. White EV (1942) J Am Chem Soc 64:2838

3 Enzymatic Hydrolysis

Enzymatic hydrolysis of cellulose accomplishes degradation of cellulose to glucose. This heterogeneous catalytic reaction is typically characterized by an insoluble reactant (cellulose) and a soluble catalyst (enzymes) [150]. The rate of this reaction is influenced by both structural features of cellulose and mode of enzyme action. Unfortunately, the enzymatic hydrolysis of native cellulose proceeds at an extremely low rate, and therefore, its pretreatment prior to hydrolysis is essential to enhance the rate of hydrolysis. This chapter elaborates on the nature of lignocellulosic structural resistance, properties and mode of enzyme action, different pretreatment methods, and a variety of kinetic models for enzymatic degradation.

3.1 Nature of Lignocellulosic Structural Resistance

The ability of cellulolytic microorganisms and that of cell free cellulolytic enzymes to degrade cellulose vary greatly with the structural features of the lignocellulosic materials. The structural features of lignocellulosic materials which govern their susceptibility to enzymatic degradation include [64]: (i) the moisture content of the fiber, (ii) the size and diffusivity of the cellulolytic enzymes and other reagent molecules involved relative to the size and surface properties of the grown capillaries, and the space between microfibrils and the cellulose molecules in the amorphous region, (iii) the degree of crystallinity of cellulose, (iv) the unit cell dimensions of cellulose, (v) the conformation and steric rigidity of the anhydroglucose units, (vi) the degree of polymerization of the cellulose, (vii) the nature of the substances with which cellulose is associated, and (viii) the nature, concentration, and distribution of substituent groups. Some of the structural features that are important in enzymatic hydrolysis of cellulose will be elaborated.

3.1.1 Degree of Water Swelling of Cellulose Fiber

The properties of wood and cellulose are profoundly affected by water. In partially dried cellulose, changes in the quantity of hygroscopically-bound water induced by variations in the relative humidity of the surrounding atmosphere govern many mechanical and physical properties and the extent of swelling or shrinkage. Little or no swelling of wood or cellulose takes place in relatively nonpolar liquids, e.g., benzene, whereas significant swelling occurs in polar solvents, e.g., water [175].

The volumetric swelling of wood in water varies from 9 to 21.1%, depending on the type of wood. The expanded capillary structure of water swollen fibers may cause a significant increase in surface area of the cellulose fiber, and the total area of the swollen material may be as much as 100-fold greater than the area that results after drying the material [36]. This enlarges the fine structure so that the substrate is more accessible to cellulolytic enzymes. This also facilitates diffusion of extracellular enzymes, and furthermore, molecules of water are added to the cellulose during hydrolytic cleavage of glycosidic links [64].

3.1.2 Crystallinity of Cellulose

An important structural feature that affects the rate of enzymatic hydrolysis of cellulose fibers is the degree of crystallinity of cellulose. The crystallinity of native cellulose was experimentally determined by Segal et al. [254] with an X-ray diffractometer using the focusing and transmission techniques. They measured the intensity of the 002 interference and the amorphous scatter at $2\theta = 18°$. The fraction of crystalline material in the total cellulose was expressed in terms of an X-ray crystallinity index. Dunlap et al. [51] has reanalyzed the data reported by Baker et al. [20] as shown in Fig. 3.1, to examine the relationship between the

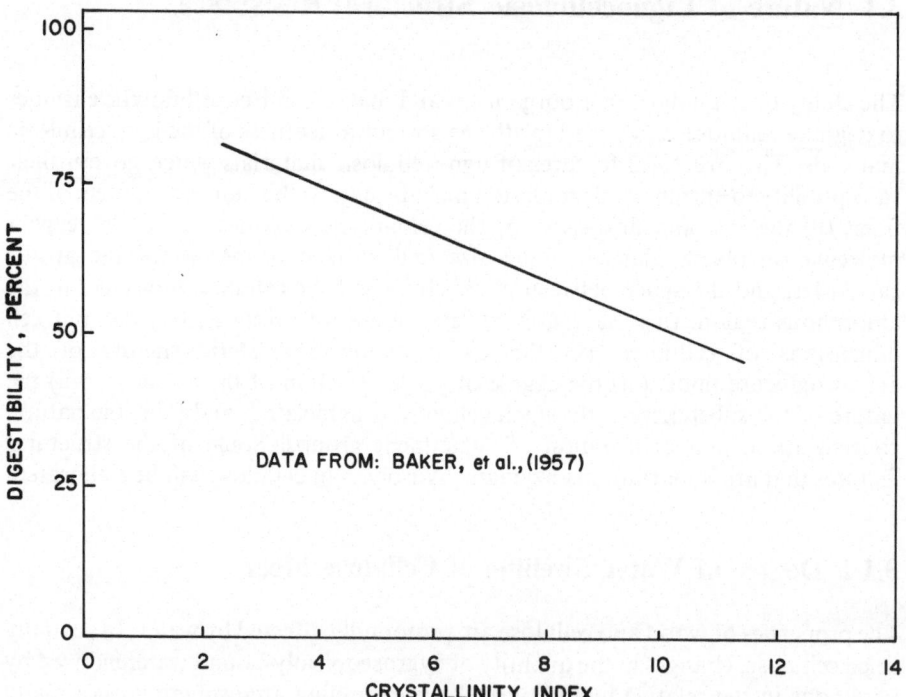

Fig. 3.1. Digestibility of cotton linters and wood cellulose against x-ray crystallinity index [51]

cellulose crystallinity and digestibility. This figure indicates an inverse linear relationship between the crystallinity index and digestibility. Norkrans [199] and Walseth [298] showed that the cellulolytic enzymes degrade the more readily accessible amorphous portions of regenerated cellulose, but are unable to attack the less accessible crystalline portions. They observed a significant increase in crystallinity during hydrolysis of cellulose with enzymes. With depletion of the highly accessible amorphous material, the substrate becomes more crystalline, thereby offering an increased resistance to further hydrolysis [241].

Caulfield and Moore [38] reported somewhat different observations from those of Reese et al. [241]. Their measurement of the degree of crystallinity of ball-milled cellulose before and after partial hydrolysis indicated that the mechanical action (ball-milling) increases the susceptibility of both the amorphous and crystalline components of cellulose. They also observed that grinding enhances the enzyme digestibility of the crystalline component to a greater extent than that of the amorphous component. They concluded that the overall increase in the digestibility is apparently a result of decreased particle size and increased surface rather than a result of reduced crystallinity.

3.1.3 Molecular Structure of Cellulose

Cellulose exists in four recognized crystal structures designated as celluloses I, II, III, and IV [110]. The sources of these four structural types are shown in Fig. 3.2. Rautela and King [230] cultivated *Trichoderma viride* on cellulosic substrates with four different crystal forms and determined the activation energy for the enzymatic hydrolysis of each separate substrate. Their results showed that this organism

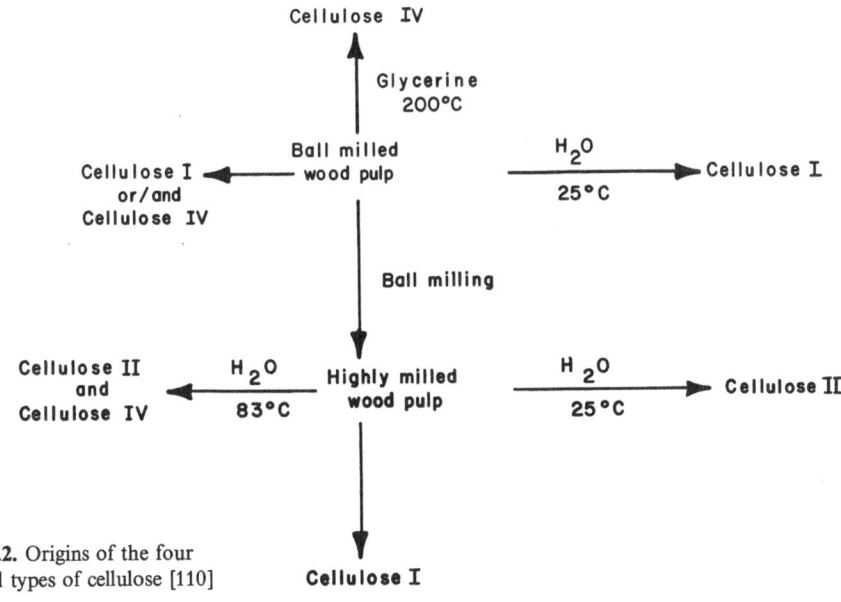

Fig. 3.2. Origins of the four crystal types of cellulose [110]

23

produces an enzyme that yields the minimum activation energy for hydrolysis of a specific substrate. In other words, this fungus can synthesize an enzyme having active sites of a unique structure, which accommodate a specific crystal lattice structure of an individual substrate.

King [130] has suggested that the greater resistance of crystalline as compared to amorphous cellulose may not be due solely to its physical inaccessibility to enzyme molecules, but also to the conformation and steric rigidity of the anhydroglucose units within the crystalline regions.

A study by Rowland [246] of the selective availability of hydroxyl groups in fibrous and crystalline cartons has shown that the relative availability of hydroxyl groups on the most highly crystalline segments of the elementary fibril approached 1, 0 and 0.5 for $O_{(2)}H$, $O_{(3)}H$, and $O_{(6)}H$, respectively. These values of the theoretical relative availability on the surface of crystalline cellulose I result from the complete unavailability of $O_{(3)}H$ for reaction as a result of hydrogen bonding to $O_{(5)}$, and 50% unavailability of $O_{(6)}H$ as a result of hydrogen bonding to $O_{(1)}$.

3.1.4 Extraneous Material – Lignin

The combination of lignin with partially crystalline cellulose existing in wood constitutes one of nature's most biologically resistant material [65]. Micro-organisms are prevented from degrading cellulose in wood by the presence of lignin. While the chemical structure of lignin is well understood, the nature of its association with the wood polysaccharides remains uncertain [300]. Three main theories are prevalent: (A) hydrogen bonding between constituents, (B) covalent chemical bonds, and (C) incrustation, thereby preventing easy access to enzyme molecules. A currently accepted view is that it is largely physical in nature, meaning that lignin and amorphous cellulose form a mutually interpenetrating system of high polymers, but some covalent links do exist between lignin and hemicellulose [300].

Since the intact lignin coating represents a great constraint to the accessibility of cellulose, the greatest need in the development of lignocellulosic bioconversion processes is for methods that will diminish the protective effect of lignin. As discussed earlier, almost all potential substrates for cellulose bioconversion are heavily lignified. Thus, most of the cellulose in nature will remain unsuitable for bioconversion unless effective and economically viable procedures are developed to remove or modify lignin [132, 244]. The essential feature of any successful pretreatment is to decrease the protective association between lignin and cellulose. It is, however, unnecessary to remove or alter all the lignin to significantly increase susceptibility to enzymatic degradation. Depending upon the source of cellulose, only 20–65% of the lignin need be removed [69, 179].

Several well developed methods are available to bring about the disintegration of the cellulose-lignin association with the aid of chemicals which decompose lignin, leaving the cellulose largely intact. Such processes are technically called pulping or digestion [37].

Conventional pulping processes, such as Kraft pulping and sulfite pulping [37], are too costly as bioconversion pretreatments; moreover, they pose a number of

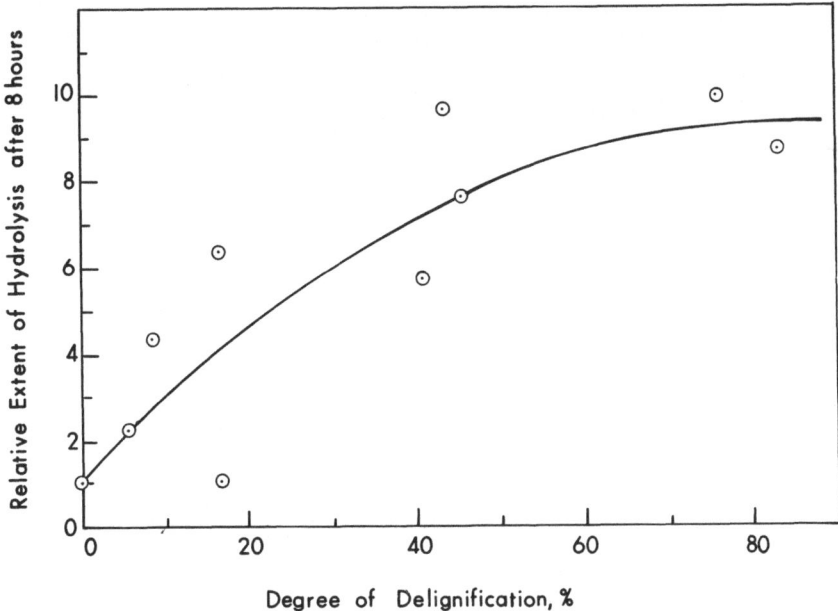

Fig. 3.3. Relationship between the extent of delignification and the hydrolysis rate [69]

problems, e. g., recovery of pulping chemicals and safe disposal of waste chemicals. However, modification of these conventional pulping processes and other pulping processes deserves consideration and should be optimized for maximum increase in digestibility with minimum cost [252].

The effects of lignin content or extent of delignification of various ligno-cellulosics on the in vitro digestibility have been studied by several investigators [19, 69, 179, 295]. The relationship between the extent of delignification and the hydrosis rate by *Trichoderma reesei* QM 9414 is shown in Fig. 3.3 [69]. Fan et al. [69] observed that the hydrolysis rate for wheat straw increased substantially with an increase in the extent of delignification up to about 50 % delignification; beyond this, however, the hydrolysis rate increased only slightly.

Van Soest [295] studied the relationship between the digestible dry matter and the percentage lignin content in dry matter of forage; the results are shown in Fig. 3.4. Of the 83 kinds of forage studied, some with very high cell-wall content (grasses) have low lignin content, thus possessing a relatively high digestibility, whereas others with low cell-wall content, (legumes) have high lignin contents, thus possessing lower-digestibility. An appreciable difference in the delignification-digestibility relationship has been reported between hardwoods and softwoods [19].

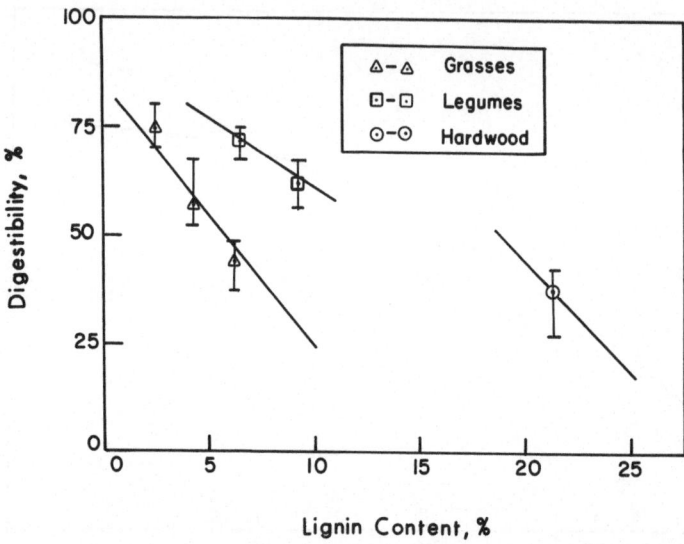

Fig. 3.4. Relationship between the dry matter digestibility and lignin content [179, 295]

3.1.5 Presence of Substituent Groups

Substituted cellulose derivatives are formed by replacing the hydrogen of the primary and secondary hydroxyl groups of cellulose with reactive groups such as methyl, ethyl, hydroxyethyl and carboxymethyl. The addition of these groups makes cellulose less crystalline and more soluble in water in proportion to the degree of substitution (DS) and the solvating capacity of the substituent groups [64]. The DS at which the complete solubility is attained ranges from 0.5 to 0.7, depending on the solvating capacity of the substituents and the degree of polymerization of the cellulose [43].

The susceptibility of substituted cellulose derivatives to enzymatic hydrolysis increases as the derivatives become more water soluble and less crystalline up to the complete solubility point. After this point, the susceptibility decreases with increasing DS until complete immunity to the enzymatic action results. This usually occurs at a DS somewhat greater than 1.0. Substituent groups of large molecular dimensions are more effective in resisting enzymatic degradation than small groups [235].

3.1.6 Capillary Structure of Cellulose

The susceptibility of cellulose to enzymatic hydrolysis is determined largely by its accessibility to cellulolytic enzymes. Direct physical contact between the enzymes and the substrate molecules of cellulose is prerequisite to hydrolysis [64]. Since cellulose is an insoluble and structurally complex substrate, this contact can be achieved only by diffusion of the enzymes into the complex structural matrix of cellulose. Any structural feature that limits the accessibility of cellulose to enzymes will diminish the susceptibility of cellulose to hydrolysis.

While a good correlation has often been obtained between a particular structural feature and the hydrolysis rate of a substrate, a great deal of ambiguity remains. The lack of agreement between different sets of data can be illustrated by the following examples. In contrast to the majority of reports, Lee [147] has found that the most reactive regenerated cellulose is also the most crystalline. This implies that the degree of swelling, not the crystallinity of the substrate, is the most important factor because a higher degree of swelling increases the pore size beyond a certain critical value which is necessary for enzyme molecules to penetrate the substrate. Another example of confusion arises in equating the accessibility of cellulose to its amorphous fraction [265]. A highly amorphous substrate may be slow to hydrolyze because the amorphous material in the wood is accessible to water and hydrogen ions but is completely inaccessible to an enzyme. These two examples emphasize the importance of capillary structure and enzyme reactivity to hydrolysis.

The capillary structure of cellulose relative to the size and diffusivity of each cellulolytic enzyme is the most significant structural feature. Direct physical contact between the enzyme and its substrate produces an enzyme-substrate complex which then breaks down to yield the reaction product. It is expected, therefore, that the rate of reaction should be a function of the surface area of cellulosic fibers as defined by the size, shape and surface properties of the microscopic capillaries within the fiber in relation to the size, shape, and diffusibility of the cellulolytic enzyme itself.

The capillary voids in cellulosic fibers include: (i) gross capillaries such as the cell lumina, pit apertures and pit-membrane pores that are visible in the light microscope and are in the range between 200 Å and 10 or more microns in diameter, and (ii) cell wall capillaries such as the spaces among microfibrils and those among cellulose molecules in the amorphous regions, which are less than 200 Å in diameter [43]. The total surface area exposed in the gross capillaries is fairly large, approximately 2×10^3 cm^2 per gram of wood or cotton. It is, however, several orders of magnitude smaller than the total surface area exposed within the cell wall capillaries, which is approximately 3×10^6 cm^2 g^{-1} [41]. The area exposed on the gross capillary surfaces of one gram of wood or cotton is sufficient to accommodate about 3×10^{15} randomly oriented enzyme molecules 200×35 Å in size, which is equivalent approximately to 3 mg of enzyme protein per gram of wood or cotton. It should be noted that the spaces among the microfibrils and cellulose molecules in the amorphous regions are occupied by hemicellulose and lignin. Cutting or grinding the wood would produce a considerable amount of additional contacting surface between the cellulose and enzyme. If reactivity is not increased by vibratory ball milling or other treatment methods, it must be due to the changes within the cell wall itself rather than due to the size of the cell wall fragment.

The dimensions of the cellulose capillaries have been determined with the solute exclusion technique by Aggebrandt and Samuelson [3], Stone and his coworkers [265], Nelson and Oliver [185], and Van Dyke [294]. Figure 3.5 shows a frequency distribution of capillary dimensions in water-swollen wood, cotton, and wood pulp. The median and maximum dimensions of cell wall capillaries in these three

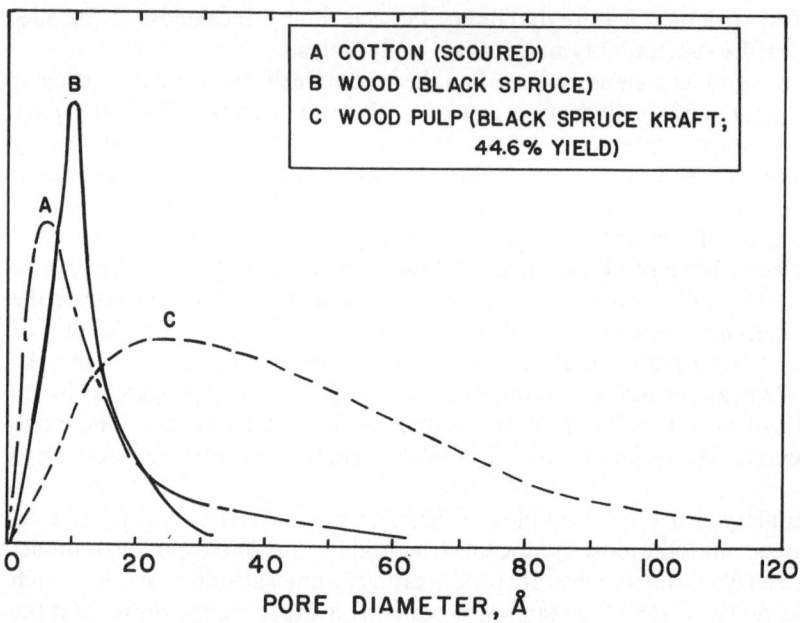

Fig. 3.5. Frequency distribution for cell wall capillaries [43]

materials are: about 5 and 40 Å for water-swollen wood; about 10 and 75 Å for water-swollen cotton, and about 45 and 110 Å for water-swollen wood pulp [43]. The substantial increase in the median and maximum pore sizes during pulping results, in part, from removal of lignin and hemicellulose from cell wall capillaries.

The dimensions of cellulolytic enzyme molecules range from 24 to 77 Å in diameter, with an average of 59 Å [42]. Thus, these molecules would be expected to diffuse readily within the gross capillaries and act on cellulose molecules exposed on the surface of these capillaries. However, only a small fraction of the cell wall capillaries in water swollen wood and cotton fibers is sufficiently large to permit penetration of most of the cellulolytic enzyme molecules. For this reason, cellulolytic enzymes are likely to be physically excluded from all but the largest cell wall capillaries. These capillaries must be enlarged to render them accessible to the enzymes. Stone et al. [265] measured the pore volume with the polymer exclusion technique and calculated the surface area within the swollen fibers accessible to all the molecules. The substrate was reacted with a commercial cellulase enzyme preparation, and the initial rate of reaction was compared with the accessibility of the substrate to molecules of various sizes. They found a linear relationship between the initial reaction rate and the surface area within the cellulose gel which was accessible to a molecule of 40 Å diameter. Figure 3.6 shows the relationship between the initial rate of reaction and the surface area of cellulose accessible to molecules of different sizes. The digestibility of cellulose with different degrees of swelling appears to be directly proportional to the accessibility of molecules of 20–40 Å in diameter, which might correspond to the size of cellulose molecules.

28

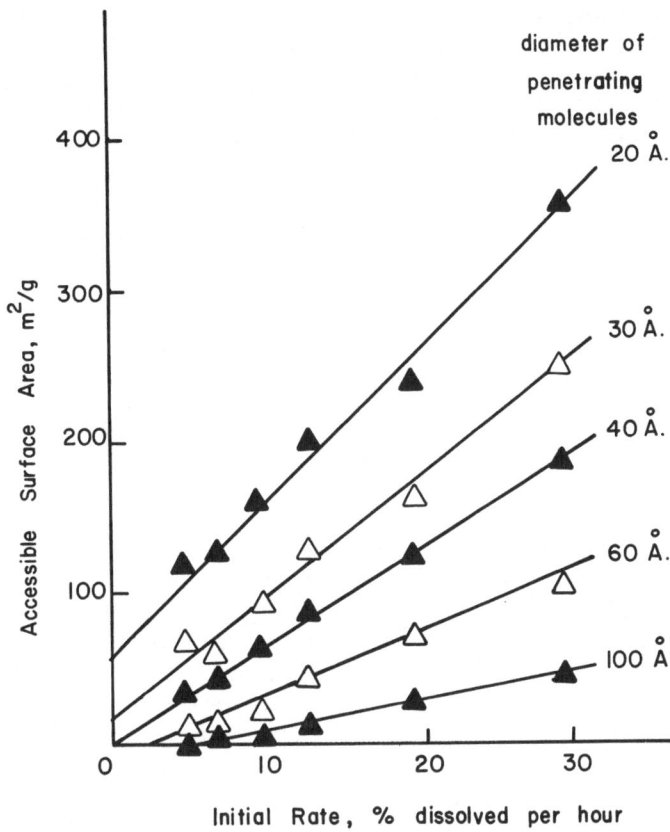

Fig. 3.6. Relationship between the initial rate of reaction and the surface area of cellulose accessible to molecules of different sizes [265]

3.2 Properties and Mode of Cellulase Biodegradation

Cellulase refers to a group of enzymes that contribute to the degradation of cellulose to glucose. The native crystalline cellulose is water insoluble, and its structure and complexity render it highly resistant to the hydrolytic action of enzymes [152]. In most cellulolytic organisms, several cellulase components form a cellulase complex which synergistically hydrolyzes cellulosic substrates. Reese and his coworkers [240] have proposed the so-called (C_1-C_x) hypothesis to explain the mechanism of cellulase action. According to this hypothesis, the reaction involves at least two enzyme components. The first component, C_1, activates or deaggregates the cellulose chains in preparation for attack by the next hydrolytic component of the cellulase complex. The presence of component C_1 is essential if highly ordered substrates are to be attacked. The second component C_x, hydrolyzes soluble derivatives of cellulose or swollen and partially degraded cellulose; however, highly ordered substrates are not attacked.

To understand the mode of action of a cellulase system, the substrate specificity of each of the purified cellulase components must be fully investigated. Prior to 1964, the microorganisms capable of extensive hydrolysis of the highly ordered forms of cellulose had not been screened [152]. Moreover, techniques for fractionating cellulase components were not well developed. As a result, most of the early investigations with cellulases were carried out using the C_x component. Several active C_x-type enzymes were found in culture filtrates of cellulolytic microorganisms. One component attacks cellulose endo-wise (randomly), and it exhibits different relative activities to different substrates. Other fractions act exo-wise (endwise) by removing successive units of glucose from the nonreducing end of a cellulose chain.

Recent studies of the mechanism of cellulase action have been focused mainly on the isolation, purification, and characterization of component C_1. Several investigators have supported the idea that component C_1 has a nonhydrolytic function, as originally proposed by Reese and his coworkers [240]. However, the results of recent fractionation studies suggest that component C_1 is not nonhydrolytic but a β-1,4-glucan cellobiohydrolase [25, 55, 93, 329]. These investigators have proposed that the $(C_1$–$C_x)$ hypothesis be abandoned and the mechanism of cellulase action be reformulated. According to their new concept, crystalline cellulose is effectively rendered soluble by the cooperative action of endo-glucanase and exo-glucanase enzymes; the latter removes cellobiose from the end of the cellulose chain. Many questions still remain unanswered and their resolution depends on the development of purification techniques for cellulase fractions.

Comprehensive accounts of many aspects of cellulase and their application are available in the literature [18, 73, 74, 88, 152, 231, 233, 261, 299]. The papers on cellulase presented at the 4th and 5th International Fermentation Symposia have been published collectively [47, 279]. Articles on cellulose published before 1959 are well reviewed in the book written by Gascoigne and Gascoigne [74]. A collective bibliographic list of 670 articles on cellulase and its application is available in the book edited by Reese [233]. Reese and Mandels [238] have reviewed enzymatic degradation of cellulose. Nisizawa [187] also published a comprehensive review of the mode of action of cellulases. Compendia of findings on enzyme mechanism associated with the C_1 component are also available [55, 327].

Accumulated papers on the mode of action of cellulase published since 1964 are listed in Table 3.1. These references are classified by research groups. Some of the references cited in Table 3.1 may appear to be obsolete or no longer relevant. Nevertheless, they are included to illustrate, to the fullest extent possible, the pattern of historical development of various mechanisms to explain the mode of action of different kinds of cellulase and the relationships among the proposed mechanisms. Most cell-free enzyme preparations from culture filtrates of cellulolytic microorganisms contain mainly the C_x component and β-glucosidase. Only a few cellulolytic microorganisms can produce a significant amount of the C_1 component. They are mostly fungi such as *Trichoderma viride*, *Trichoderma koningii*, and *Fusarium solani*. Consequently, the mode of action of cellulase from these strains has been studied intensively. In addition to the mode of action of cellulase, physical properties of cellulase and the effect of cellulase on some properties of cellulose fibers are considered.

30

Table 3.1. Mechanism and mode of action of cellulase [152]

Authors	Microorganisms	Fractionation methods	Comments
Mandels and Reese (1964)	*T. viride*, *Myrothecium verrucaria*	Zone electrophoresis, DEAE-dextran chromatography	1) The hydrolytic (C_x) and solubilizing (C_1) factors of the cellulase complex were partially separated 2) Synergistic action of these two components on cotton digestion was observed
Selby and Maitland (1965)	*Myrothecium verrucaria*	Gel filtration on Sephadex G-75	1) Three major cellulolytic components were fractionated 2) One component was active on CMC degradation, the other two were mainly responsible for cotton degradation 3) There was no evidence of synergism between these components
Selby and Maitland (1967)	*T. viride*	Gel filtration on Sephadex G-75, Ion-exchange chromatography on DEAE and SE-Sephadex	1) Three components were fractionated 2) The components, which are essential for attack on cotton, were a CMCase, a cellobiase, and a third C_1 component which had no action on CMC, cellobiose, or cotton 3) Synergism was observed among these components
Selby (1969)	*T. viride*, *Penicillium funiculosum*	Gel filtration on Sephadex G-75, Ion-exchange chromatography on DEAE and SE-Sephadex	1) C_1 components, which have very similar properties, were isolated from *Trichoderma* and *Penicillium* 2) Synergism and cross-synergism between C_1 and C_x components were demonstrated for cotton degradation
Li et al. (1965)	*T. viride*	Chromatography on Avicel column	1) β-1, 4-D-glucan glucohydrolase (EC 3.2.1.21), β-1, 4-D-glucan glucanohydrolase (3.2.1.4), and an enzyme component (C_1), capable of attacking crystalline cellulose, were purified 2) Synergism was observed among these components on cotton digestion 3) Distinct physical and enzymatic properties of each component were observed 4) C_1 component has a nonhydrolytic function
Liu and King (1967)	*T. viride*	Not indicated	1) When Avicel was treated with partially purified C_1-cellulase, a marked increase in the total number of particles occurred during the initial stage of reaction 2) C_1 component may not be an enzyme but rather a protein, which would cause the initial particles to gradually collapse and release the ultimate micelles

31

Table 3.1. (continued)

Authors	Microorganisms	Fractionation methods	Comments
Halliwell (1965)	*Trichoderma koningii*	Cell free filtrates	1) The early enzymic breakdown of cotton fibers is characterized by the formation of very short fibers
Halliwell and Riaz (1970)	*Trichoderma koningii*	Sephadex G-75, DEAE-Sephadex, CM-Sephadex, Cellulose powder column	1) Two CMCase components and a C_1 component were separated 2) The C_1 component acted weak and only on cotton, forming soluble products but not short fibers 3) The ability to form short fibers was confined to one of the CMCases, and its action was unaffected by the other components
Halliwell and Riaz (1971)	*T. koningii*	Similar to 1970 paper	1) Four apparently pure components, cellobiase, C_1 like component, CMCase, and that named component C_2, were isolated 2) CMCase showed mainly short fiber-forming activity 3) Interaction of these four components in the degradation of native cellulose was discussed 4) Synergism was observed between C_2 and CMCase
Halliwell and Griffin (1973)	*T. koningii*	Recycling chromatography on DEAE-Sephadex and on Sephadex G-75	1) C_1 component was isolated, and it was shown to act as a β-1,4-glucan cellobiohydrolase 2) This component released terminal cellobiose units from cellulose. C_x component was not required for the action of C_1, however, the enzyme synergized extensively with cellobiase 3) The relationship between this C_1 component and the entire cellulase complex was discussed
Halliwell and Griffin (1974)	*T. koningii*	Not indicated	1) The activity of a highly purified C_1 component was shown to be that of an exocellulase, which liberates a cellobiose unit from both native and simple substrates. This reaction was promoted not by C_x, but by cellobiose 2) C_x were composed of two components: CMCase and another component capable of hydrolyzing cellulose, but not CMC
Wood (1968)	*T. koningii*	Sephadex G-75, DEAE-Sephadex, SE-Sephadex	1) C_1 component was isolated 2) This component had little ability to produce soluble sugar from cotton; but when recombined with other components, this capacity was almost recovered 3) C_1 component had no swelling factor activity on its own, but it had a synergistic effect on the swelling factor activity with other

Reference	Organism	Methods	Results
Wood (1969)	*Fusarium solani*, *T. koningii*	Ion exchange on DEAE-Sephadex, Gel chromatography on Sephadex G-100	1) C_1 components of *F. solani* and *T. koningii* were similar in physical properties 2) C_1 components of each strain synergized with the other strains C_x fraction (cross synergism)
Wood (1971)	*Fusarium solani*	Heat treatment, Gel filtration on Sephadex C-100, Iso-electric focusing	1) The β-1,4-glucanase and the β-glucosidase components were purified 2) The specificity of these two components was studied
Wood (1972)	*T. koningii*, *F. solani*	Gel filtration on Sephadex G-75 or G-100, Ion-exchange chromatography on DEAE-Sephadex, Electrofocusing	1) C_1 component was separated. This C_1 component possessed a slight ability to produce reducing sugar from CMC 2) Strong synergism was observed between C_1 and C_x components 3) Cellobiose was the sole product of C_1 action on Walseth, dewaxed cotton, cellohexaose, and cellotetraose 4) C_1 is cellobiosylhydrolase
Wood and McCrae (1972)	*T. koningii*	DEAE-Sephadex, with a salt gradient and pH gradient, Electrofocusing	1) C_1 component was isolated 2) This component had little ability to attack CMC and native cellulose, but degraded Walseth cellulose readily and dewaxed cotton to the extent of 15% 3) C_1 component is a β-1,4-glucan cellobiosylhydrolase
Wood and McCrae (1975)	*T. koningii*	A series of fractionation and purification procedures	1) Eight highly purified components were purified, and all of them were found to be hydrolytic enzymes 2) C_1 acted synergistically to a great extent with more random acting C_x components
Wood and McCrae (1977)	*Fusarium solani*	Ultrogel AcA-54, DEAE-Sephadex,	1) The purified C_1 component showed little capacity for hydrolyzing highly ordered substrate, but hydrolyzed readily Walseth soluble cellooligosaccharides 2) C_1 is cellobiohydrolase
Emert et al. (1974)	*T. viride*	Avicel adsorption column, DEAE-Sephadex A-50, Preparative disc gel electrophoresis	1) Three forms of cellobiohydrolase and a cellobiase were purified 2) The function of the cellobiohydrolase is cleavage of cellooligosaccharides at the site of their production (i.e., the cellulose surface) 3) Inhibition of this component by the cellobiose was observed
Gum and Brown, Jr. (1976)	*T. viride*	Ultrafiltration using Amicon PM 30, Avicel column, DEAE-Sephadex A-50	1) A glycoprotein enzyme β-1,4-D-glucan cellobiohydrolase (E.C. 3.2.1.91) was purified 2) Physicochemical properties were determined

Table 3.1. (continued)

Authors	Microorganisms	Fractionation methods	Comments
Petterson et al. (1972)	*T. viride*	Gel filtration on Bio-Gel P-10, Ion exchange on DEAE-Sephadex A-50, Preparative gel electrophoresis in polyacrylamide gel	1) C_1 component was isolated 2) C_1 component is a hydrolytic enzyme, it functions according to an endwise mechanism, possibly by removing cellobiose units 3) The synergistic effect with C_x enzymes was explained by supposing that the C_x enzymes randomly split β-1,4 glucosidic bonds and that the C_1 enzyme consecutively removes cellobiose units from the free end
Berghem and Pettersson (1973)	*T. viride*	Chromatography on Bio-Gel P-10, DEAE-Sephadex chromatography, Isoelectric focusing, chromatography on Bio-Gel P-60	1) Chemical and physicochemical properties of C_1 component were determined 2) Avicel, Walseth, and cellotetraose were degraded by the enzyme, and in each case the product was cellobiose 3) C_1 component is a β-1,4-glucan cellobiohydrolase
Pettersson (1975)	*T. viride*	Bio-Gel P-10, Ion-Exchange chromatography, Isoelectric focusing, Bio-Gel P-60	1) Four cellulolytic enzymes were purified 2) One of the enzymes was an exo-β-1,4-glucanase, which catalyzes the hydrolysis of microcrystalline cellulase up to 80% solubilization
Berghem et al. (1975)	*T. viride*	Same as 1972 paper	1) β-1,4-glucan cellobiohydrolase was further characterized as regards chemical, physicochemical, and enzyme properties 2) The enzyme is a β-1,4-glucan cellobiohydrolase 3) This enzyme can degrade microcrystalline cellulose up to 80% within 72 h, if the cellobiose is continuously removed
Berghem et al. (1976)	*T. viride*	Molecular sieve chromatography, Dipolar adsorbent chromatography, Isoelectric focusing Affinity chromatography	1) A low molecular weight and a high molecular weight β-1,4-glucan glucanohydrolase (C_x) were isolated 2) Protein composition, molecular weight were determined 3) Both enzymes were active in releasing free fibers from filter paper, but low molecular weight enzyme was more effective
Eriksson and Rzedowski (1969)	*Chrysosporium lignorum*	DEAE-Sephadex A-50 column	1) Three cellulase peaks were separated 2) Strong indication for the existence of a C_1 enzyme was found

Reference	Organism	Method	Results
Eriksson and Pettersson (1975a)	*Sporotrichum pulverulentum*	Fractionation on DEAE-Sephadex and Gel filtration and SP-Sephadex, Concanavalin A-Sepharose, Polyacrylamide-gel electrophoresis, Isoelectric focusing on flatbed polyacrylamide gel	1) Five pure endo-β-1,4-glucanases were separated 2) One exo-glucanase was identified 3) Endo-glucanases were purified and physicochemical characterizations were determined
Eriksson and Pettersson (1975b)	*Sporotrichum pulverulentum*	DEAE-Sephadex, Gel filteration Activation on Dowex 2-X8 anion exchanger, Concanavalin-A-Sepharose, SP-Sephadex	1) An exo-β-1,4-glucanase was purified 2) Physicochemical properties were determined
Almin et al. (1975)	*Sporotrichum pulverulentum*	Same as Eriksson and Pettersson (1975)	1) Five types of purified endo-β-1,4-glucanase were characterized with regards to their molecular activities and Michaelis Menten constants
Eriksson (1975)	*Sporotrichum pulverulentum*	Not indicated	1) The synergistic action among five types of endo-β-1,4-glucanases and an exo-β-1,4-glucanase was demonstrated 2) Quinone oxidoreductase was found
Iwasaki et al. (1964)	*T. koningii*	DEAE-Sephadex A-25, Amberlite IRC-50 (XE-64) Hydroxylapatite column	1) Two types of cellulase, cellulase I and II, were separated 2) Cellulase I had higher activity toward glycol cellulose and cellobiose, but no measurable activity toward insoluble fibrous cellulosic material 3) Cellulose II could easily decompose fibrous cellulose, but did not act on glycol cellulose and cellobiose 4) Both enzymes were homogeneous, and had different properties
Ogawa and Toyama (1964)	*T. viride, Aspergillus niger*	Cellulose column using pulverized filter paper or disintegrated gauze or cellulose powder	1) Three cellulase components were proposed: C_1 (native cellulose hydrolyzing enzyme), C_1 (acting on filter paper of swollen cellulose), C_3 (active on CMC) 2) Avicel was adopted as a new substrate for determination of activity
Ogawa and Toyama (1966)	*T. viride, A. niger*	Gauze column, DEAE-Sephadex A-50, Electrophoresis	1) Crude fractionation of cellulase from *T. viride* and *A. niger* was carried out 2) A cellulolytic component capable of degrading filter paper was separated 3) Cellulase components, which are active on Avicel, gauze, filter paper, and CMC, were studied

Table 3.1. (continued)

Authors	Microorganisms	Fractionation methods	Comments
Ogawa and Toyama (1967)	*T. viride*	Gauze column, DEAE-Sephadex A-50	1) Four cellulase activities: the gauze degrading (C_1), Avicel decomposing (C_{1a}), filter paper degrading (C_2) and CMC decomposing (C_3) activities, were separated 2) Synergism among these components was observed
Ogawa and Toyama (1968)	*T. viride*	Gauze column method	1) Cellulolytic enzymes were separated into nonadsorbed and adsorbed fractions by the gauze column method 2) The enzyme in the nonadsorbed fraction included cellobiase, CMCase, and Avicelase activity, but neither filter paper nor gauze degrading activities. Adsorbed fraction contained CMCase and Avicelase 3) Synergism was observed between these two fractions
Ogawa and Toyama (1972)	*T. viride*	Gauze column method, DEAE-Sephadex A-50, Amberlite CG-50, Sephadex G-100, Gel filtration	1) Nonadsorbed and adsorbed fractions were further purified 2) Nonadsorbed fraction was composed of two components, and consisted of mainly CMCase activity. This fraction revealed no activities capable of degrading filter paper but showed trace activity against Avicel 3) Adsorbed fraction had three components which were active on filter paper, Avicel, and CMC. This fraction was considered to be an essential of natural cellulose decomposition
Watanabe (1968a)	*Chaetomium globosum*	Cellulose powder column, DEAE-Sephadex A-25, Amberlite XE-64	1) Three cellulase fractions were separated 2) The enzyme system consisted of at least three fractions
Watanabe (1968b)	*Chaetomium globosum*	Cellulose powder, DEAE-Sephadex A-25, Amberlite XE-64	1) One cellulase fraction, which acted on CMC, was separated 2) Properties of this fraction were determined
Niwa et al. (1964)	*T. viride*	Starch zone, Electrophoresis, Hydroxylapatite, DEAE-Sephadex	1) Cellulase contained 3–4 isozymes, some exhibited strong CMCase activity and others were active on filter paper 2) Synergism was observed among these components 3) Use of various substrates in measuring enzyme activity was suggested

Reference	Source	Method	Findings
Niwa et al. (1965)	*T. viride*	Column chromatography using Amberlite, DEAE-Sephadex, Electrophoresis	1) Seven peaks of cellulase fractions were purified 2) One component was highly active on CMC, but much less active on insoluble substrates such as cotton and filter paper 3) One component, which showed homogeneous electrophoretic pattern, exhibited the C_1 and C_x activities 4) Further separation of this C_1 and C_x components was attempted, but was not successful
Nisizawa et al. (1966)	*T. viride*	Amberlite CG-50 column	1) Study on the nature and mechanism of swelling factor was performed 2) Swelling factor activity was probably involved in the cellulase activity itself 3) Semicrystalline region in absorbent cotton was suggested
Okada et al. (1966)	*T. viride*	Amberlite CG-50, DEAE-Sephadex A-50	1) Five cellulase components were separated 2) One cellulase component hydrolyzed insoluble substrates, e.g., filter paper, more easily than CMC, cellodextrin and several cellooligosaccharides 3) It also showed a swelling factor activity toward absorbent cotton. Thus this cellulase possesses C_1 as well as C_x activity with a random mechanism
Okada et al. (1968)	*T. viride*	Amberlite CG-50 DEAE-Sephadex A-50	1) Three types of β-1,4-glucan glucanohydrolase were obtained 2) The properties of each component were examined 3) These types of cellulases hydrolyzed a series of substrates, including several cellooligosaccharides, CMC, and various insoluble celluloses such as Avicel and cotton, but they showed no activites toward cellobiose
Toda et al. (1968)	*T. viride*	Amberlite CG-50, DEAE-Sephadex A-50, and Sephadex G-75 column chromatography	1) Kinetic constants of three purified cellulase components for various cellooligosaccharides were evaluated
Tomita et al. (1968)	*T. viride*	Amberlite CG-50 column chromatography, DEAE-Sephadex A-50 column chromatography	1) Chromatographic patterns of cellulase components of *T. viride* grown on the synthetic and natural media were compared with those of commercial product Meicelase and cellulase Onozuka
Nisizawa et al. (1972)	*T. viride*, *Irpex lacteus*	Amberlite CG-50 column, DEAE-Sephadex A-50 column chromatography, Isoelectric focusing method	1) Two components, Avicelase and CMCase, were separated 2) Avicelase exhibited less random mechanism on attacking cotton fiber and CMC 3) CMCase reduced the degree of polymerization of cotton fiber more rapidly than Avicelase, but produced a small amount of reducing sugar 4) Synergism was observed between two components of both strains 5) The authors supported the C_1–C_x concept, but also proposed that their individual function should be modified

Table 3.1. (continued)

Authors	Microorganisms	Fractionation methods	Comments
Tomita et al. (1974)	*T. viride*	Amberlite CG-50, DEAE-Sephadex A-50, Biogel P-150	1) Further purification of Avicelase was attempted 2) Avicelase attacked Avicel very readily to produce a mixture of cellobiose and a few percent glucose, but attacked CMC less readily 3) Avicelase also attacked cotton and Walseth at a lower rate than Avicel 4) Avicelase was regarded as a cellulase component of the random type but it was shown to be much less random
Okada (1975)	*T. viride*	DEAE-Sephadex A-50, Sephadex G-100, Sephadex G-75, Starch column chromatography	1) Two cellulase components were separated 2) Some properties of these cellulase were investigated
Okada and Nisizawa (1975)	*T. viride*	Same as Okada (1975)	1) Two highly purified cellulases were obtained 2) Both cellulases produced predominantly cellobiose and glucose 3) Both cellulases were able to perform transglycosylations
Okada (1976)	*T. viride*	Similar to Okada (1975)	1) A cellulase component was purified 2) The enzyme was characterized as a less random type with regards to its action on CMC, and produced predominantly cellobiose and glucose from various cellulosic substrates as well as from higher cellooligosaccharides
Kanda et al. (1976a)	*Irpex lacteus*	Amberlite CG-50, Bio-gel P-100 gel, DEAE-Sephadex A-25, Sephadex G-100 gel Starch column, Sephadex G-100 gel rechromatography	1) An endo-cellulase was purified, and its physical properties were determined 2) This endo-cellulase attacked a series of cellooligosaccharides, β-cellobioside, CMC, and insoluble cellulosic substrates 3) Cellobiose was the largest product from digestion of insoluble substrates
Kanda et al. (1976b)	*Irpex lacteus*	Not indicated	1) Three endo-cellulase components were separated; one was CMCase and the others were of the Avicelase group 2) A mixture of CMCase and Avicelase group gave a remarkable synergistic action in degradation of crystalline cellulose

3.2.1 Effect of Cellulase on some Properties of Cellulose Fibers

Prior to producing a measurable quantity of reducing sugar, native cellulose undergoing attack by cellulase exhibits extensive changes in physical properties, such as transverse cracking, considerable loss in tensile strength, lowering of the degree of polymerization, increased capacity for moisture uptake and for alkali absorption, and fragmentation to short separable fibers [140, 187, 238, 256]. These changes are associated with the insoluble nature of cellulose substrates and occur simultaneously or successively. The most significant phenomena among them are the fragmentation and swelling of native cellulose.

3.2.1.1 Fragmentation of Cellulose Fibers

Marsh [169] reported that transverse cracks were observed to occur in cotton fibers upon a short exposure to cellulolytic enzymes, and he theorized that the enhanced fragility of cotton fibers may be due to these transverse cracks. Notable investigations on the fragmentation of cellulose fibers have been carried out. Halliwell [92] reported that the early enzymatic breakdown of cotton fibers is characterized by the formation of a large number of very short fibers that increases to a maximum and then decreases gradually by conversion to soluble sugars. The enzyme from $T. koningii$ converted a minor fraction (up to 16%) of substrate into soluble sugar and a major portion (80%) into insoluble short fibers within 20 h. These fragments were converted further into the reducing sugar. Marsh [168] also reported that little production of soluble material occurred during the fragmentation of cotton fibers. The loss of fiber as nonsedimentable solid material did not exceed 6%, while 75% of the starting material was converted to shorter fibers during the treatment of dewaxed cotton fibers with a cellulase solution.

A similar observation on fragmentation was obtained by King [128, 129], Rautela and King [230] and Liu and King [159] for various forms of crystalline cellulose treated with a crude cellulase preparation from $T. viride$. Rautela and King [230] compared electronmicrographs of undegraded and 65% crystalline cellulose I. The electron microscopic observation revealed that the enzymatic action resulted in fissures parallel to the long axes of the crystallite aggregates and in production of many fine particles of similar shape and size.

Ogiwara and Arai [206] observed the attack of $T. viride$ cellulase on bleached sulfite pulp and compared the action of the cellulase with the action of acid. Their work indicates that like the acid, the cellulase preferentially attacks the amorphous regions of cellulose fibers. However, they found that the morphological character of cellulose fiber changed entirely upon treatment with acid. Several researchers attempted to identify the enzyme fraction involved in this fragmentation. Based on the C_1–C_x hypothesis, the early stages of attack on cotton fibers can be tentatively identified with the C_1 action [237]. The C_1 component of $T. viride$ studied by Selby and Maitland [259] was also active in producing these short fibers. Liu and King [159] found that the C_1 component would induce the initial particles to gradually collapse and release the ultimate miscelles. Wood [323] reported similar results.

The fragmentation was studied in some depth by Halliwell and Riaz [95, 96]. They separated two CMCase and two C_1 components and found that the separated

C_1 component acts weakly on cotton fibers, forming a soluble product, but does not contribute to short fiber formation. The short fiber forming activity was attributed only to enzymes showing C_x activity, and their action was unaffected by the other components. Wood [327] further purified the enzyme fraction and found that fragmentation of the Avicel is characteristic of the purified C_x enzyme action rather than the C_1 enzyme action. Pettersson [223] also reported that the free fiber-forming activity is caused by endo-glucanase.

3.2.1.2 Swelling of Cellulose Fibers and the Swelling Factor

The most rapid and sensitive measurement of activity of cellulolytic enzymes is provided by the alkali swelling test. In an early work, Marsh et al. [170] found that fibers invaded by some fungi or those treated with their culture filtrates enhance the absorbability of 18% NaOH. They assume that this phenomenon is due to the action of an enzyme-like entity produced by the fungi, and they named it the "swelling factor" (S-Factor). Later, Reese and Gilligan [237] made a further study of this S-Factor in an attempt to fit it into the general action pattern of cellulase. They confirmed that its behavior is that of an enzyme, concluded that the S-Factor could probably be included with the C_x enzymes, and suggested that swelling must occur as a result of the partial damage of the primary wall by the attack of the swelling factor secreted from the microorganism. This swelling occurs before release of reducing sugar, loss in weight, or decrease in the degree of polymerization.

On the other hand, according to Youatt and Jermyn [334] and Youatt [333], swelling is caused by the synergistic effect of several cellulases acting together to attack native cellulose and that the swelling effect itself is partially a consequence of the supermolecular architecture of a given fiber. Based on electron microscopic observations, Norkrans [198] has postulated that the extent of swelling after enzymatic attack may depend on the condition of fibrils which constitute the network structure of the primary and secondary layers in cotton fibers. A similar view was presented by Selby [256], who has concluded that the swelling of cotton fibers is enhanced by the loosening of the constricting effect of the winding layer during hydrolysis by cellulase.

Later Nisizawa et al. [192] investigated the swelling factor of Meicelase, a commercial cellulase preparation from *T. viride*, using dewaxed absorbent cotton and mercerized cotton. They observed that these cellulose samples showed a similar response to the swelling factor, and the degree of swelling increased very rapidly especially at the early stage of incubation. The average degree of polymerization of cotton decreased in parallel with an increase in the extent of swelling during incubation with a cellulase preparation. In view of these results, Nisizawa et al. [192] have postulated that besides the crystalline and amorphous regions another region must exist in the microfibrils of cotton fibers, namely, a semicrystalline region with a moderate crystallinity. This region can easily absorb alkali even when only a few β-1,4-glucosidic bonds therein are cleaved by cellulase. However, this region is unable to fully absorb alkali in the ordinary state but that following the attack by cellulase it might change temporarily to an amorphous state which can absorb more alkali.

40

3.2.2 Mode of Action of the Cellulase

3.2.2.1 (C_1–C_x) Concept

Reese et al. [240] first suggested a mechanism for the enzymatic breakdown of cellulose involving the C_1 component. The conversion of native cellulose to soluble sugar was depicted as a two-step process as follows: The C_1 component "activates" or deaggregates cellulose chains so that the enzymes classified as C_x can carry out the depolymerization as shown in Fig. 3.7. Microorganisms growing only on soluble forms of cellulose, e. g., carboxymethylcellulose (CMC), can synthesize only the C_x component, whereas microorganisms capable of growing on highly ordered forms of cellulose produce both C_1 and C_x. However, this postulate of Reese et al. [240] is yet to be confirmed. They, as well as other investigators, found that the cell-free culture filtrates show high activity only toward soluble, swollen, and partially degraded cellulose, which is hydrolyzed by the C_x enzymes. Because of the difficulty involved in preparing filtrates active against highly ordered forms of cellulose, e. g., native cotton, much of the work in the period, 1950–1964, was related to the studies of the C_x activity. The discovery in 1964–1965 that culture filtrates prepared from certain strains of *Trichoderma viride* and *Trichoderma koningii* are capable of extensive hydrolysis of native cellulose was a turning point in the study of cellulase. An extensive search for the C_1 component was followed. During this period, the separation of the C_1-like components from the culture filtrate of *T. viride* and *T. koningii* was reported by several groups of workers [92, 117, 155, 163, 194, 195, 201].

Numerous reports dealing with the fractionation of the culture filtrates have appeared since 1964, and, in general, these have stated that cellulase, as secreted by *T. viride* and *T. koningii*, is a multicomponent enzyme system. Substantial evidence has been found to support the existence of the C_1 component; nevertheless, it appears that the majority of the investigators did not carry out the purification steps to completion, and thus their views differ widely with regard to the specificities of the various C_1 enzymes that have been isolated. More recently, culture filtrates of *Fusarium solani* [324], and *Penicillium funiculosum* [257] have been shown to be highly active on cotton, and these have also been fractionated.

Several investigators have supported the idea that the C_1 component has a non-hydrolytic function. Mandels and Reese [163] have further developed the (C_1–C_x) hypothesis on the basis that a culture filtrate of *T. viride* chromatographed on a

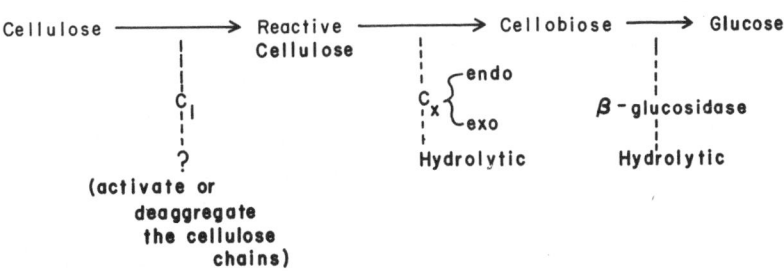

Fig. 3.7. Degradation of cellulose; the (C_1 – C_x) concept [163]

DEAE-dextran column was observed to separate into several components including the C_1 and C_x components. Increases in the reducing sugar from cotton silver and CMC were used as measures for the activities of C_1, and C_x, respectively. It was suggested that cotton silver is first solubilized by C_1, thereby changing it into a form similar to that of swollen or generated cellulose, which, in turn, is hydrolyzed by C_x into soluble fragments. The C_x component shows no activity on cotton, but does show activity on hydrolyzed amorphous cellulose as well as on CMC. The idea that the C_1 component has a nonhydrolytic function was supported by Li et al. [155].

Li et al. [155] purified three cellulase components: an enzyme removing successive glucosyl moieties from the nonreducing end, an enzyme attacking internal bonds, and a component capable of attacking crystalline cellulose. They postulated that the C_1 component has a nonhydrolytic function. Later Liu and King [159] observed that when Avicel was treated with a partially purified C_1 component, a marked increase in the total number of particles occurred during the initial stage of reaction. They postulated that the C_1 component may not be an enzyme but rather a protein capable of causing the initial particles to collapse gradually and release the ultimate miscelles.

Isolation of the C_1 component as an enzyme entity was reported by Selby and Maitland [258, 259] and Selby [255]. This component was first found in the culture filtrates of *T. viride* and then in other sources such as *P. funiculosum*. The C_1 component becomes active only when the CMCase fraction from the same culture filtrate is added to it, but the C_1 component by itself shows no cellulolytic capacity. Selby and Maitland [258, 259] and Selby [255] concluded that this component represents a true C_1. Furthermore, they postulated two possible explanations for the synergistic action of C_1 and C_x. The first points out that the C_1 component makes the cotton fiber more susceptible to the action of the C_x component. The second postulates that C_1 is inhibited by its own reaction products until they are removed by the action of C_x.

King and Vessal [131] have summarized the cellulase complex based on the (C_1-C_x) hypothesis as follows: C_1 is an enzyme whose action is unspecified but is required for the hydrolysis of highly oriented solid cellulose by β-1,4-glucanase $(= C_x)$. C_x is a group of hydrolytic enzymes that exist as two types: (a) exo-β-1,4-glucanase, which successively removes individual glucose units from the non-reducing ends of cellulose chains, and (b) endo-β-1,4-glucanase, which randomly attacks internal linkages that are more susceptible to hydrolysis than terminal linkages. According to the (C_1-C_x) theory, cotton fibers are attacked successfully by C_1-C_x, and β-glucosidase (cellobiase) and are finally converted into glucose.

On the other hand, some experimental results appear to contradict the (C_1-C_x) theory. The C_1 component, which is the essential component in hydrolyzing crystalline cellulose, has shown a hydrolytic rather than a nonhydrolytic function [62, 95, 190, 203–205, 286, 323]. Ogawa and Toyama [201, 204, 205] reported on the separation of two cellulase components. One component consisted mainly of CMCase activity and revealed no activities capable of degrading filter paper but showed trace activity against Avicel. The other component was active on filter paper, Avicel, absorbent cotton, as well as CMC. Thus, these researchers considered the latter to be the essential component of the natural cellulose decomposing activity of *T. viride*. Wood [323] separated a C_1 component that had little ability to produce soluble sugar from cotton but when recombined with other components almost totally recovered this capacity. In addition, this C_1 component had no swelling factor activity by itself, but it had a synergistic effect on the swelling factor activity with other fractions. Strong indication for the existence of a C_1 enzyme from another strain, *Chrysosporium lignorum*, was also reported [62].

Experimental results by Okada et al. [210, 211] also appear to contradict the $(C_1–C_x)$ theory. These investigators obtained three highly purified cellulase components from *T. viride*, each of them showing both C_1 and C_x activities; that is they hydrolyzed randomly not only cotton fiber and Avicel but also CMC and various β-1,4-oligoglucosides excluding cellobiose, although the differences are considerable for the relative ratios of the activities of the three highly purified enzyme components for each of these substrates. All purified cellulase components appear to be of the mixed $(C_1–C_x)$ type, notwithstanding the fact that the C_x activity in one component may be different from that in another with each exhibiting a unique substrate specificity.

Nisizawa et al. [193] further purified the cellulase complex and separated two different components, namely, Avicelase and CMCase. The purified CMCase easily and randomly saccharifies highly amorphous cellulose, CMC, and cellooligosaccharides. It produces glucose and cellobiose in almost equal amounts; however, it does not readily attack Avicel and cotton. In contrast, the purified Avicelase splits not only highly crystalline cellulose such as Avicel and cotton but also cellooligosaccharides to produce primarily cellobiose from these substrates. Avicelase also hydrolyzes CMC, but in a much less random fashion than CMCase. Therefore, Avicelase seems to cut off one cellobiosyl residue at a time, mainly from the end of these substrates. Nisizawa et al. [193] also found that a mixture of CMCase and Avicelase shows a remarkable synergistic effect toward highly crystalline cellulose such as Avicel and cotton, but not highly amorphous cellulose such as phosphoric acid swollen cellulose [298] and CMC. A similar result was obtained from a culture filtrate of *Irpex lacteus* [193]; thus, according to Nisizawa et al. [193], CMCase and Avicelase from these fungi seemed to correspond to C_1 and C_x, but it must be CMCase that attacks cellulose initially to yield cellulose fragments with low degrees of polymerization. These fragments, in turn, serve as substrates for Avicelase; this cellulase saccharifies the cellulose fragments from their ends. Therefore, they have suggested that the roles of C_1 and C_x are opposite to the roles originally proposed by Reese et al. [240]. Nisizawa et al. [193] have also reported that the separated C_1 component is a hydrolytic enzyme. This component functions according to an endo-wise mechanism, possibly by removing cellobiose units. The synergistic effect with the C_x enzyme is attributed to the C_x enzymes randomly splitting β-1,4-glucosidic bonds and to the C_1 enzyme consecutively removing cellobiose units from the free end.

3.2.2.2 β-1,4-Glucan Cellobiohydrolase

Some investigators have supported the concept that the C_1 component has a nonhydrolytic function. However, based on the results of recent fractionation studies, several investigators have suggested that the concept be abandoned and that the mechanism of cellulase action be reformulated. The new concept postulates that crystalline cellulose is effectively rendered soluble by the cooperative action of an endo-glucanase enzyme and an exo-glucanase enzyme, which perform the special function of removing cellobiose from the end of the cellulose chain.

The properties and mode of action of an individual enzyme of the cellulase complex can be described with confidence only when its purification is complete. Chromatography on columns of Avicel, hydroxylapatite, Amberlite ion-exchange resins, alkali swollen cellulose, and degraded cotton gauze have been used to fractionate cellulolytic enzymes (see Table 3.1). The most successful results have been obtained with fractionation using a combination of two or more packing materials. Some researchers have effectively used a combined procedure of gel filtration and chromatography on various forms of ion-exchange Sephadex in the fractionation of the cellulase complex.

Wood [328] and Wood and McCrae [329, 331] reported on the isolation and purification of the C_1 component produced by *T. koningii* and *Fusarium solani*. The isolation and purification were carried out by gel filtration on Sephadex G-75 or G-100, ion exchange chromatography on DEAE-Sephadex, and electrofocusing in a stabilized pH gradient spanning over half on pH unit. The C_1 component obtained still possessed a limited ability to produce reducing sugar from a solution of CMC but showed no capacity for reducing the viscosity of the solution. It was concluded that the C_1 activity and a trace of C_x (CMCase) activity originated in the same enzyme protein. Furthermore, the C_1 component, acting alone, was unable to attack highly ordered celluloses; by acting synergistically with the separated C_x and β-glucosidase components, it was able to accomplish extensive breakdown. Cellobiose was virtually the sole product of the action of C_1 on (a) phosphoric acid-swollen cellulose, (b) dewaxed cotton fiber that had been 15% degraded by exposure to a highly concentrated enzyme for four weeks, and (c) cellohexaose and cellotetraose. Wood [328] and Wood and McCrae [329, 331] have advocated that the C_1 component is a β-1,4-glucan cellobiohydrolase. Wood and McRae [327] have postulated that the C_1 is a special type of hydrolytic enzyme which is unable to attack CMC or crystalline cellulose to any significant extent; nevertheless, this enzyme possesses the capacity for degrading cellulose substrates in a very accessible form by removing cellobiose residues successively from the chain ends. The cellobiohydrolase was also separated from *Fusarium solani* by Wood and McCrae [330].

Halliwell and Griffin [93, 94] have confirmed that C_1 acts as a β-1,4-glucan cellobiohydrolase. They isolated a C_1 component from *T. koningii* cellulase which was free from the CMCase activity but could still solubilize cotton to an extent of 20% in 7 days when acting along. When this C_1 component was mixed with a cellobiase component having no activity on cotton, the degree of solubilization was increased to 35%. Extending the period of incubation to 21 days resulted in a marked increase in the solubilization (to 70%) when the cellobiase was contained, but no increase in the solubilization activity when only C_1 was contained. The conclusion was that the cellobiose liberated by the C_1 action acts as an inhibitor.

Using the combined procedure of gel filtration, ion exchange, and gel electrophoresis, Pettersson et al. [224] separated the C_1 component from *T. viride*. They found that the C_1 component is a hydrolytic enzyme that removes cellobiose units according to an exo-wise mechanism. Therefore the conclusion has been drawn that the C_1 component is a β-1,4-glucan cellobiohydrolase. Later Berghem and Pettersson [25] and Berghem et al. [26] further purified this component and characterized the chemical and physicochemical properties of its subcomponents. Emert et al. [55] also reported the separation of three forms of cellobiohydrolase and one cellobiase and proposed that the function of the cellobiohydrolases is cleavage of cellooligosaccharides.

The evidence cited by the various authors to support the claim that C_1 is a cellobiohydrolase appears to be convincing. In spite of this, striking differences remain in the extent to which highly ordered cellulose is hydrolyzed by the various C_1-type components, even when the enzyme is almost free from a contaminating CMCase activity.

3.2.2.3 C_x Components

The C_x complex of enzymes consists of the hydrolytic enzymes that hydrolyze β-1,4-bonds in a cellulose molecule. The designation covers the random-acting endo-β-1,4-glucanase and endwise acting exo-β-1,4-glucanase. With the assistance of C_1, the β-1,4-glucanases hydrolyze crystalline cellulose; in the absence of C_1, they hydrolyze only "noncrystalline" cellulose such as that produced by swelling, grinding, or reprecipitation from a solution [238]. The activity of the β-1,4-glucanases is usually measured by the action on soluble cellulose derivatives, e.g., carboxymethylcellulose (CMC). The subscript x in C_x emphasizes the multicomponent nature of the fraction containing C_x enzymes. The β-1,4-glucanases (C_x) clearly exist as two types: (a) endo-β-1,4-glucanase with action of a random nature, the terminal linkages generally being less susceptible to hydrolysis than the internal linkages, and (b) exo-β-1,4-glucanase, successively removing single glucose units from the nonreducing end of the cellulose chain. Enzyme commission numbers have been assigned to some of the cellulase components; endo-β-1,4-glucanase is E.C. 3.2.1.4, and β-glucosidase is E.C. 2.2.1.21 [131]. The latter designation does not differentiate between arylglucosidase and dimerase (cellobiase). The exo-β-1,4-glucanase has sometimes been placed in E.C. 2.2.1.21, but a new number should be assigned to it [131].

Endo-β-1,4-Glucanase. Many researchers carried out experiments to determine if the glucose chains of cellulose are degraded by cellulase according to an endo-wise or an exo-wise mechanism. Karrer et al. [122] and Nisizawa and Kobayashi [188, 189] initially theorized that cellulase might hydrolyze the long chains of cellulose molecules according to the exo-wise mechanism. Subsequent evidence indicates that both mechanisms are active [74]. Although Storvick and King [266] later proposed an exo-wise mechanism for hydrolysis by the cellulases of *Cellvibrio gilvus*, other works such as those of Kooiman et al. [139], Reese [234], and Hash and King [106] suggested that the attack by several cellulases on polymeric substrates is basically random.

Cellulases of the following microorganisms act randomly: *Myrothecium verrucaria* [305, 307, 309, 310, 312], *Irpex lacteus* [119, 120, 191], *Chaetomium globosum* [302, 303], *Polyporus versicolor* [220, 222], *Sporotrichum pulverulentum* [5, 60, 61], *Cellvibrio gilvus* [267], *Pseudomonas fluorescens* [190, 273], *Penicillium notatum* [218, 219, 221], *Fusarium solani* [325, 326], *Trichoderma koningii* [92, 326]. Cellulolytic enzymes of endo-β-1,4-glucanase from the culture filtrate of *Trichoderma viride* have been isolated and investigated extensively [155, 163, 195, 201–205, 207, 208, 210, 259, 280].

The role of endo-glucanases in the overall process of enzymatic cellulose degradation is that they split the cellulose chains in less-ordered regions of the cellulose fiber, thus creating starting points for attack by exo-glucanases [27, 329]. However, the exact role of the endoglucanases in the degradation of cellulose fibers is not yet fully understood.

Reese and Mandels [238] have summarized the mode of action of endo-glucanases as follows: The end-enzymes are the characteristically random-acting glucanases responsible for hydrolysis of the high-molecular-weight glucans. Too few ends exist in these glucans to obtain appreciable rates of hydrolysis by exo-enzymes. These enzymes act preferentially on longer chains. Acting on the soluble cellulose derivatives, these random acting enzymes cause a rapid decrease in viscosity, with a relatively slow increase in reducing end groups. The action is not entirely random, because both end linkages are somewhat less affected than the internal linkages. These endoglucanases can also hydrolyze the soluble cellodextrins. The rate of hydrolysis increases with the degree of polymerization within the limits of substrate solubility. Cellobiose is hydrolyzed by most endo-glucanases, but not by all. The major products of the endo-glucanase action are cellobiose and cellotriose.

Carboxymethylcellulase. It is a class of C_x which acts on the degradation of soluble cellulose derivatives, especially CMC. For convenience it is called "CMCase" [192]. The effect of chemical modification of cellulose on its susceptibility to endo-glucanase action is complex. Any action taken to increase the solubility of a polymer in water or to open the crystalline structure will increase the rate of hydrolysis. Simultaneously the presence of a substituent effectively prevents hydrolysis of linkages adjacent to it. In other words, the extent of hydrolysis is reduced as the number of substituents increases [193].

The enzymatic hydrolysis of CMC usually results in a loss of viscosity and an increase in the reducing power of CMC. When the rate of the viscosity reduction is larger than the rate of the reducing power increase, a "more" exo-wise hydrolysis is expected to predominate. Based on this observation, Gilligan and Reese [81] identified cellulase components with different degrees of randomness in the hydrolysis of CMC. They showed that these cellulase components coexist even in a single microorganism. Later, similar observations were made on several micro-organisms, such as *Irpex lacteus*, *Trichoderma viride*, and *Pseudomonas fluorescens*, by Nisizawa et al. [190, 191] and Okada et al. [210, 211]. These workers have separated several cellulase components with different rates of CMC liquefying and saccharifying activities. When increases in specific viscosities are plotted against reducing power, the slopes of the resultant curves are unique for each component. Components whose curves are similar to those obtained from an acid hydrolysis are considered "more" randomly acting. These results suggest that several cellulase components exist in each organism and that they are different in their detailed or fine specificity toward CMC [280].

Exo-β-1,4-Glucan Glucohydrolase. Whitaker [309] has reported that swollen cellulose is hydrolyzed by *Myrothecium* cellulase by an exo-wise mechanism. His conclusion has been drawn from loss in weight of the substrate, decrease in the number-average degree of polymerization, and decrease in the weight-average degree of polymerization during incubation. However, many investigators have suggested that the attack by several cellulases on a polymeric substrate is basically random. The work of Storvick et al. [267] has demonstrated that an exo-wise cellulase component may exist in the culture filtrate of *Cellvibrio gilvus*. Li et al. [155], King [126], and King and Vessal [131] have obtained a similar component from *T. viride* cellulase. Since this enzyme component degrades cellodextrins and CMC in an exo-wise fashion, it has been termed exo-glucanase. This enzyme however, attacks insoluble cellulose with difficulty, and it is presumed to be a type of CMCase.

According to Reese [232, 236] and Barras et al. [21], many exo-β-1,4-glucanases exist in various microorganisms in nature. These enzymes have been investigated less frequently than endo-glucanases. In general, exo-β-1,4-glucanases successively remove glucose units from nonreducing ends of the glucan. They act most rapidly on the soluble chains of four to seven units produced by the action of an endo-enzyme and, as such, play an important part in the overall digestion of cellulose to glucose. Because of their high specificity, components of enzymes have very limited action on cellulose derivatives and on glucans having mixed linkages or branches [190]. Although they can remove one or two unsubstituted terminal units, progress down the chain is halted by any deviation from the regular structure.

3.2.2.4 β-Glucosidase

The mode of action of β-glucosidases has been summarized by King and Vessal [131]. β-glucosidases vary in specificities, and the ones that act primarily on aryl-β-glucosides are not involved in the cellulase action. The β-glucosidases in cellulose breakdown are those that hydrolyze cellobiose and other very short-chain β-1,4-oligoglucosides to glucose, especially those highly active on the β-dimers of glucose including cellobiose. Cellobiase is the most descriptive designation for the cellulase complex system, but a more accurate name might be "β-glucimerase", indicating the ability to act on all of the β-dimers of glucose. These enzymes do not act on insoluble cellulose. A comparison of the mode of action of β-glucosidase and that of the exo-α-glucanase reveals the following [239]. β-glucosidase usually acts on substrates ranging from cellobiose to cellohexaose. Both exo-glucanases and β-glucosidase act on cellobiose, cellotriose, and cellotetraose. The exo-enzyme acts preferentially and by inversion on longer oligomers. β-glucosidase hydrolyzes dimers and trimers very rapidly, but the rate decreases markedly as the degree of polymerization increases. The β-configuration is retained by the product. Furthermore, β-glucosidase is strongly inhibited by gluconolactone, and it hydrolyzes (1→1)-β, (1→2)-β, and (1→6)-β as well as (1→4)-β linkages. Exo-glucanases have relatively high linkage specificity. The enzyme commission number assigned to β-glucosidase is E.C. 3.2.1.21; however, the designation does not differentiate between arylglucosidase and dimerase (cellobiase) [131].

Gong et al. [84] separated three distinct cellobiase components from a commercial *T. viride* cellulase preparation by repeated chromatography on DEAE cellulose eluting with a salt gradient. An evaluation of the physical properties, the kinetics, and the mechanism of these components indicates that the components are subject to product inhibition and that they hydrolyze cellobiose by a noncompetitive mechanism.

3.2.2.5 Synergism Among Components

It is now well established that cellulase is a multicomponent enzyme complex and that crystalline cellulose is hydrolyzed by synergism of these cellulase components (see Table 3.1). The synergism has been discussed by many investigators. In this section, we shall focus on synergism of the C_1 component (cellobiohydrolase) with other components.

Wood [327, 328] separated C_1 and C_x components from *T. koningii* cellulase on DEAE-Sephadex. When each component was diluted to make its concentration equivalent to that of the original culture filtrate from which it was separated, the fraction containing the C_x activity retained only 3% of the cellulase activity of the original culture filtrate, and C_1 component only 4%. However, when the two fractions were recombined in their original proportions, 96% of the cellulase activity of the original culture filtrate was recovered. Clearly, the activity of the cellulase complex was dependent on the synergistic action of the C_1 component with other components (see Table 3.2).

Halliwell and Griffin [93] fractionated enzyme fractions from the same microorganism as that investigated by Wood [327, 328] and studied their synergism. There was limited synergism between the C_1 component and other components of the cellulase complex, but when all the components in the complex were recombined in their original proportions, most of the activity of the original culture filtrate was recovered.

Suga et al. [271] theoretically analyzed the degradation of polysaccharides by endo- and exo-enzymes. They have considered the degradation of polysaccharide chains by endo-enzymes alone and by various combinations of endo- and exo-enzymes. The latter case is related to cultures of *T. viride* in that these cultures have different exo- and endo-enzymes.

Table 3.2. Relative cellulase activities of the components of *T. koningii* cellulase [327]

Enzyme	Recovery of cellulase activity, %
$C_x + \beta$-glucosidase	3
C_1	4
$C_1 + C_x + \beta$-glucosidase	96
Original culture filtrate	100

The derivation of the model of Suga et al. [271] is outlined below. The reaction between the enzyme and substrate is assumed to be of the common Michaelis-Menten type.

$$S + E \underset{k_2}{\overset{k_1}{\rightleftharpoons}} ES \overset{k_3}{\longrightarrow} P + E. \tag{3.1}$$

This gives rise to the following kinetic equation;

$$\frac{dC_i}{dt} = -k_1 C_E (i-1) C_i + k_2 (C_{ES})_i + 2k_3 \sum_{j=i+1}^{\infty} \{(C_{ES})_j / (j-1)\}. \tag{3.2}$$

Specifically, for the enzyme-glycosidic bond complex, we have

$$\frac{d(C_{ES})_i}{dt} = k_1 C_E (i-1) C_i - (k_2 + k_3)(C_{ES})_i \tag{3.3}$$

and for the free enzyme, we have

$$C_E = C_{Et} - \sum_{i=2}^{\infty} (C_{ES})_i. \tag{3.4}$$

Imposing the quasi steady-state approximation, we obtain from Eq. (3.3)

$$k_1 \left\{ C_{Et} - \sum_{i=2}^{\infty} (C_{ES})_i \right\} (i-1) C_i - (k_2 + k_3)(C_{ES})_i = 0. \tag{3.5}$$

Solving the above equations gives

$$\frac{dC_i}{dt} = -\frac{k_3 C_{Et} (i-1) C_i}{K_m + \sum\limits_{j=2}^{\infty} (j-1) C_j} + 2 \frac{k_3 C_{Et} \sum\limits_{j=i+1}^{\infty} C_j}{K_m + \sum\limits_{j=2}^{\infty} (j-1) C_j}. \tag{3.6}$$

When only the exo-enzyme is present, the rate of change of C_1 is given by

$$\frac{dC_1}{dt} = -\frac{k_3' C_{Et}' C_i}{K_m' + \sum\limits_{j=2}^{\infty} C_j} + \frac{k_3' C_{Et}' C_{i+1}}{K_m' + \sum\limits_{j=2}^{\infty} C_j}. \tag{3.7}$$

In the case of the degradation of a cellulose fragment of chain length i by various combinations of endo- and exo-enzymes, we obtain

$$\frac{dC_i}{dt} = -\frac{k_3 C_{Et} C_i (i-1)}{K_m + \sum\limits_{j=2}^{\infty} (j-1) C_j} + 2 \frac{k_3 C_{Et} \sum\limits_{j=i+1}^{\infty} C_j}{K_m + \sum\limits_{j=2}^{\infty} (j-1) C_j}$$

$$- \frac{k_3' C_{Et}' C_i}{K_m' + \sum\limits_{j=2}^{\infty} C_j} + \frac{k_3' C_{Et}' C_{i+1}}{K_m' + \sum\limits_{j=2}^{\infty} C_j}. \tag{3.8}$$

From the solution of these kinetic equations, the degradation index (D.I.) can be defined as

$$\text{D.I.} \equiv \frac{\mu_n(0)}{\mu_n(t)} - 1, \tag{3.9}$$

where $\mu_n(0)$ and $\mu_n(t)$ are the number average molecular weights at time $= 0$ and t, respectively. Figure 3.8 shows the numerical solution for the case of $K_m = 0.925 \times 10^{-6}$ mole of glucosidic bond ml^{-1} and $S = 10^{-3}$ g ml^{-1}. Note that the exo-enzyme by itself gives a lower rate because of the limited number of substrate sites available to it, even though its concentration is 10 times higher than that of the endo-enzyme. Again, the overall rate of action of the combined enzymes is higher than the sum of the rates of the individual enzymes.

The model of Suga et al. [271] does not consider branch points of linkages, inhibition by products, or diffusion and adsorption effects between enzymes and substrates. Nevertheless, the effect of the synergistic action of endo- and exo-enzymes can be evaluated on the basis of their model.

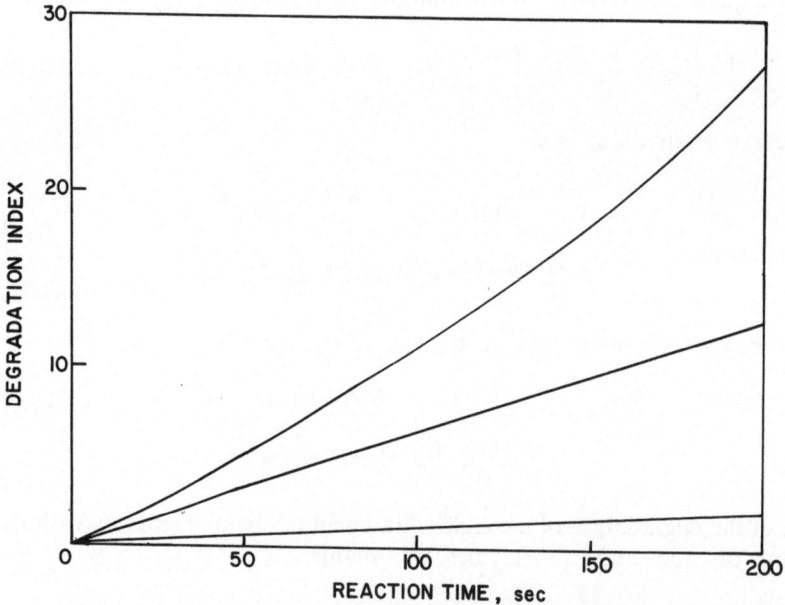

Fig. 3.8. Rate of change of the degradation index for exo-enzyme only (*lower curve*), endo-enzyme only (*middle curve*), and a combination of both enzymes at these concentrations. Degradation index is given in Eq. (3.9) [271]

3.2.3 Physical Properties of Cellulase

3.2.3.1 Molecular Weights of Cellulase Components

Molecular weights of various enzyme components are summarized in Table 3.3. The molecular weight of C_1 component from *Trichoderma sp.* lies in the range of 53,000–62,000. The endo-β-1,4-glucanases (including C_x) show considerable variation in size, with molecular weights ranging from 5,300 to 55,000. The smaller particles possibly represent subunits of a normal enzyme. Except for the smaller particles, the molecular weights of most of the enzymes range from 40,000 to 50,000. Molecular weights of 50,000–76,000 have been reported for exo-glucanase and β-glucosidase. Ahlgren and Eriksson [4] have suggested that molecules of the glucosidase are generally larger than those of endo-glucanase. Very few estimates of the size of β-glucosidase molecules have been made. Wood [324] has obtained a molecular weight of 40,000 for the *F. solani* cellulase. The molecular weight of β-1,4-glucan cellobiohydrolase isolated from culture filtrate of *T. viride* was measured at approximately 42,000 [26].

Table 3.3. Estimated molecular weights of cellulase components of selected microorganisms [152]

Micro-organism	C_1 component		C_x component		β-Glucosidase	Ref.
	C_1	CBH or Avicelase	Endoglucanase	Exoglucanase		
Trichoderma viride	60,000		52,000	76,000		[155]
	57,000		13,000			[259]
	53,000		44,000			[210]
			45,000			[208]
		46,000				[25]
		42,000				[26]
			12,500			[27]
			– 50,000			
		48,400				[87]
					76,000	[84]
Trichoderma koningii	62,000		13,000			[323]
			26,000			[259]
			– 51,000			
Fusarium solani	45,000		37,000	40,000		[324]
					40,000	[326]
Penicillium notatum			35,000			[4]
			34,500			[219]
			– 35,500			
Penicillium funiculosum	57,000					[257]
Sporotrichum pulverulentum			28,300	48,600		[60, 61]
			– 37,500			
Myrothecium verrnearia			5,300			[258]
			30,000			
			55,000			
			49,000			[311]

3.2.3.2 Diffusion Coefficient and Sedimentation Constant of Cellulase

Whitaker and his coworkers [311] were apparently the first ones who attempted to determine the diffusion coefficient and sedimentation constant of a cellulase isolated from culture filtrates of the mold *Myrothecium verrucaria*. The results were as follows: the diffusion coefficient, $D_{w,20}$, $(5.6 \pm 0.1) \times 10^{-7}$; the sedimentation constant, $S_{w,20}$, $(3.7 \pm 0.05) \times 10^{-13}$; the fractional coefficient, f/f_0, which is the frictional factor of the hydrodynamically equivalent sphere, 1.44 ± 0.02.

3.2.3.3 Size and Shape of Cellulase Molecules

The most complete physicochemical characterization of a cellulolytic enzyme molecule was carried out by Whitaker et al. [311] on the cellulase of *Myrothecium verrucaria*. Their data indicate that this particular enzyme is an asymmetric globular protein with a molecular weight of about 63,000. The molecule is considered to be cigar-shaped, roughly 200 Å long and 33 Å wide at the widest point. The validity of this judgment about the molecule's shape is supported by comparison of hydrodynamic parameters and estimates of the size and shape derived from the available viscosity and diffusion data for well characterized globular proteins, serum albumin and trytomysin, which are elongated rod shaped proteins.

Cowling [42] has estimated the sizes and dimensions of the cellulolytic enzymes of various microorganisms (Table 3.4). In some cases the dimensions given have been calculated by the procedure of Laurent and Killander [145] from the available gel filtration data on the enzymes. In other cases the dimensions have been obtained by interpolating a log-log plot of molecular radius against molecular weight of the cellulase of *Myrothecium verrucaria* and other well-characterized globular proteins using Tanford's data [275]. As illustrated in Table 3.4, dimensions of the different cellulase preparations vary considerably. If the enzymes are spherical, they range from about 25 to 80 Å in diameter with an average of 59 Å. If the enzymes are ellipsoids with an axial ratio of about 6, they range from about 15 to 40 Å in width and from 80 to 250 Å in length, giving rise to an average size of 33×200 Å.

3.3 Pretreatment Methods

High-order molecular packing of cellulose in its crystalline regions is known to hinder the heterogeneous chemical reactions on the external surface of crystallites; in addition, the structure of lignocellulosics in the cell wall resembles that of a reinforced concrete pillar with cellulose fibers being the metal rods and lignin the natural cement. Biodegradation of untreated native lignocellulosics is very slow, giving rise to the low extent of degradation, often under 20 % [51, 64–66, 69, 75]. This low rate and extent of conversion inhibit the development of an economically feasible hydrolysis process. To increase the susceptibility of cellulosic material, structural modification by means of various pretreatment schemes is essential.

Many different pretreatments have been attempted, and the literature on this subject is voluminous. Several review articles on this subject are available [50, 118,

52

Table 3.4. Estimated size of cellulase calculated from available gel filtration (GF) and molecular weight (MW) data [42]

Organism	Mol.-wt	Equivalent		Method of estimation
		Sphere diameter (Å)	Ellipsoid W × L (Å)	
Aspergillus niger	–	58	32 × 192	GF
Chrysosporium lignorum	–	37	20 × 120	GF
		38	15 × 90	GF
Fomes annosus	–	42	23 × 140	GF
		38	21 × 125	GF
Myrothecium verrucaria	63,000	77	42 × 252	MW
Myrothecium verrucaria	49,000	63	35 × 208	MW
Myrothecium verrucaria	55,000	68	37 × 224	MW
	30,000	51	28 × 168	MW
	30,000	56	31 × 186	GF
	5,300	24	13 × 79	MW
Penicillium notatum	35,000	64	35 × 210	GF
Penicillium notatum	35,000	55	30 × 182	MW
Polyporus versicolor	51,000	64	35 × 210	MW
	11,400	33	18 × 108	MW
Stereum sanguinolentum	20,500	62	34 × 204	GF
	–	43	24 × 142	MW
Stereum sanguinolentum	–	37	20 × 120	GF
Trichoderma koningii	50,000	64	35 × 210	MW
	26,000	46	25 × 152	MW
Trichoderma viride	76,000	76	42 × 250	MW
	49,000	63	35 × 208	MW
Trichoderma viride	61,000	70	38 × 230	MW
Trichoderma viride				
(Rohm and Hass	–	30	17 × 100	GF
Cellulase 36)	–	26	14 × 85	GF
Average for all enzymes taken together		59	33 × 200	

158, 179]. Of the many pretreatment methods, some have been demonstrated to be effective in disrupting the lignin-carbohydrate complex, and others in disrupting the highly ordered cellulose structure itself. Table 3.5 summarizes the various pretreatment methods that can enhance the cellulose digestibility. These pretreatments are broadly classified as physical pretreatments, chemical pretreatments, and biological pretreatments according to their principle mode of action on the substrate. Some processes are combinations of two or more pretreatment techniques applied in parallel or in sequence.

3.3.1 Physical Pretreatments

Physical pretreatments can be classified into two general categories, mechanical and non-mechanical pretreatments. Physical forces applied in mechanical pretreatments can subdivide lignocellulosic material into fine particles which are

Table 3.5. Methods used for pretreatment of lignocellulosics [65]

Physical	Chemical	Biological
Ball-milling	Alkali	Fungi
Two-roll milling	Sodium hydroxide	
Hammer milling	Ammonia	
Colloid milling	Ammonium sulfite	
Vibro energy milling	Acid	
High pressure steaming	Sulfuric acid	
Extrusion	Hydrochloric acid	
Expansion	Phosphoric acid	
Pyrolysis	Gas	
High energy radiation	Chlorine dioxide	
	Nitrogen dioxide	
	Sulfur dioxide	
	Oxidizing agents	
	Hydrogen peroxide	
	Ozone	
	Cellulose solvents	
	Cadoxen	
	CMCS	
	Solvent extraction of lignin	
	Ethanol-water extraction	
	Benzene-ethanol extraction	
	Ethylene glycol extraction	
	Butanol-water extraction	
	Swelling agents	

highly susceptible to acid or enzymatic hydrolysis. The smaller particles have a large surface-to-volume ratio, thus rendering cellulose more accessible to hydrolysis. Mechanical grinding also causes reduction in crystallinity [68, 69]. Non-mechanical physical pretreatments cause decomposition of lignocellulosics by exposing them to harsh external forces other than mechanical forces. These pretreatments include irradiation, high pressure steaming, pyrolysis, and microwave treatment.

3.3.1.1 Milling and Grinding

Ball milling. Ball milling of lignocellulosics and cellulose is an effective method for enhancing enzymatic hydrolysis. The shearing and compressive forces of the ball milling cause a reduction in crystallinity, a decrease in the mean degree of polymerization, an increase in bulk density, and a decrease in particle size. Ball milled material also allows for a high slurry concentration, thereby reducing the reactor volume and hence, the capital cost.

Several investigators [11–13, 90, 110, 124, 164, 165, 178, 181, 200, 264, 314, 319] have studied the effect of ball milling on the digestibility of lignocellulosics or cellulose. The progress of hydrolysis for ball-milled newspaper is presented in Fig. 3.9 [165]. The figure shows the change in the saccharification rate by different ball-milling times on newsprint. Wilke and Yang [318] reported a high hydrolysis conversion of 72.9% for 325 mesh ball milled newsprint in 48 hours. Millett et al.

54

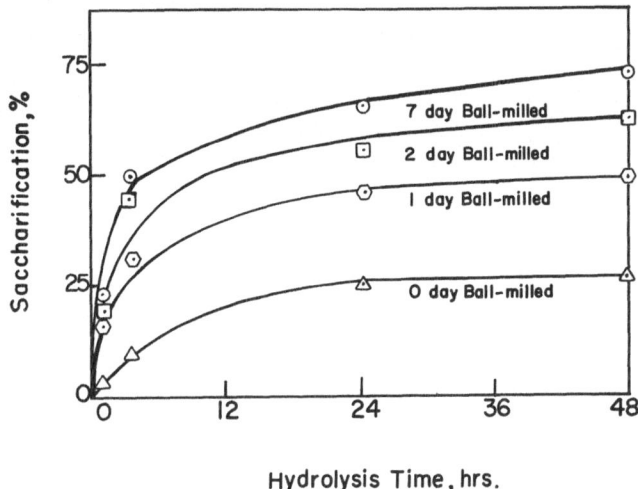

Fig. 3.9. Progress of hydrolysis for ball-milled newspaper [165]

[179] noted that the effectiveness of ball milling varied from material to material. Softwoods showed the least response; this severely limits the applicability of ball milling. Fan et al. [68] ball milled Solka Floc for a period up to 96 h. They found a marked decrease in the crystallinity of treated Solka Floc but did not observe the expected increase in the surface area. They also found that the extent of hydrolysis after 8 h was proportional to the ball-milling time (see Table 3.6). Although ball milling is an effective pretreatment method, it may be impractical on a large scale because of its lengthy pretreatment time and processing cost.

Two-roll Milling. Two-roll milling is commonly practiced in the rubber and plastics industries for grinding raw materials [278], and it has been applied to ligno-cellulosics to increase their digestibility. The mill consists of two tempered cast-iron surface rolls placed horizontally; the roll clearance can be adjusted by screws. Lignocellulosic materials are fed into the roll and masticated for a specific period of time. The pretreated material is then scraped off.

Tassinari and Macy [277] tested two-roll-milling on a wide variety of cellulosic substrates. Their findings indicated that two-roll-milled maple chips yielded 17 times more reducing sugar than the untreated maple. On the other hand, two-roll milled newspaper showed a 2.5 fold increase over ball-milled newspaper. It was also found that the sedimentation volume is lower for two-roll-milled newspaper than ball milled newspaper. This allows for a higher slurry concentration in the hydrolysis vessel, thereby reducing the reactor volume and lowering the capital costs.

Two-roll milling decreases the crystallinity and the degree of polymerization; nevertheless, its effects on lignin are not well understood. Factors that control the susceptibility to enzymatic attack are the clearance between the mill rolls and the processing time. According to Spano et al. [264], as the clearance between the rolls

55

Table 3.6. Structural parameters and extent of hydrolysis of cellulose pretreated by various methods [a] [67]

Sample	Specific surface area (SSA) ($m^2 g^{-1}$)		Crystallinity index (CrI)		Extent of hydrolysis after 8 h ($g l^{-1}$)	
	Non-solvent-dried	Solvent-dried	Non-solvent-dried	Solvent dried		
Standard substrate (Solka Floc)	2.13	3.90	74.2	77.4	14.5	14.5
Ball milled 12 h	2.09	1.54	66.3	65.1	16.8	16.8
Ball milled 24 h	2.26	1.59	30.8	59.4	18.8	18.8
Ball milled 48 h	2.36	1.59	18.7	58.1	23.3	23.3
Ball milled 96 h	1.91	1.15	4.9	36.5	30.7	30.7
γ-radiation 2 Mrad (Co cell)	–	5.37	68.1	74.1	10.1	9.9
γ-radiation 5 Mrad (Co cell)	–	4.25	75.3	77.5	9.3	8.9
γ-radiation 10 Mrad (Co cell)	–	5.50	71.3	76.0	10.8	9.6
γ-radiation 20 Mrad (Co cell)	–	7.94	72.1	75.9	12.4	10.6
γ-radiation 50 Mrad (Co cell)	–	12.80	71.1	73.4	18.1	14.3
γ-radiation 100 Mrad (reactor)	–	4.07	72.1	75.9	25.2	21.3
γ-radiation 500 Mrad (reactor)	–	106.2	64.2	70.2	59.5	37.0
Pyrolysis in air, 170°C	–	2.07	75.9	74.9	18.2	10.8
Pyrolysis in He, 170°C	–	3.61	71.0	74.6	31.6	26.2
NaOH, 1% room temperature	–	12.18	–	75.5	19.9	19.9
NaOH, 1% autoclaved	–	23.16	–	75.9	21.0	21.0
CMCS, room temperature	–	27.40	–	79.3	20.9	20.9
Concentrated acid, H_2SO_4	–	29.21	–	74.3	20.4	20.4
Microcrystalline cellulose [b]	1.84	1.97	84.5	88.8	6.2	6.2

[a] Hydrolysis conditions.
5% w/v substrate suspension at 50°C and pH 4.8.
[b] Sigmacell 50.

decreases and the processing time increases, the susceptibility of a substrate to hydrolysis increases.

Ryu et al. [247] studied the changes in cellulose structure by compression milling. They measured the crystallinity index, accessibility, and moisture recovery. The decrease in the crystallinity index was substantial and the increase in the accessibility of cellulose was drastic.

Hammer milling. A hammer mill is comprised of a rotor with a set of hammers attached. As the rotor turns, the hammers impact the substrate against a breaker plate. Hammer milling has been employed for lignocellulosic pretreatment [13, 98, 165]. Mandels et al. [165] measured the effect of hammer milling on the digestibility of newsprint and found a slight increase in susceptibility to hammer-milled newsprint over untreated newsprint. One unexpected finding was that prolonged hammer milling may actually reduce the susceptibility of cellulose.

Colloid milling. A colloid mill consists of two disks set close to each other revolving in opposite directions while the substrate slurry is passed between the disks [264]. Pretreatment by means of colloid milling has been attempted [165, 200]. Mandels et al. [165] obtained modest improvements in susceptibility of cellulose. They have concluded, however, that high operational cost makes this pretreatment uneconomical on a large scale.

Vibro energy milling. Vibro energy milling is similar to ball milling except that the mill is vibrated instead of being rotated. Vibro energy milling provides effective means for size reduction and increase in digestibility of lignocellulosics [46, 165, 178, 180, 183, 268]. Pew [225] observed that vibratory milling of spruce and aspen wood remarkably enhances their susceptibility to enzymatic hydrolysis. Similar results were obtained by Pew and Weyna [226] on spruce and aspen sawdust. Millett et al. [181] attained 63.9% saccharification upon 24 h of hydrolysis of newsprint that was Sweco-milled for six days. Sweco milling is species selective in that hardwoods are somewhat more affected than softwoods [180].

Ghose [76] achieved a 1.7 fold increase in reducing sugar for Solka Floc Sweco milled between 24 to 48 h over untreated Solka Floc. A larger increase in the reducing sugar was obtained when the substrate was heated to 200 °C before or after the pretreatment.

3.3.1.2 Pyrolysis

Pyrolysis has been investigated as a process to increase the susceptibility of cellulosic material to hydrolysis [262]. Above 300 °C, cellulose rapidly decomposes to produce gaseous and tarry compounds which leave a small amount of char residue upon evaporation. At intermediate temperatures, however, decomposition proceeds slowly and relatively few volatile products are formed.

Shafizadeh [262] reported that tar yield from vacuum pyrolysis of cellulose in a fluidized bed was 70% at 375 °C, and that at 425 °C was 78%. Mild acid hydrolysis ($1 N H_2SO_4$ at 97 °C for 2.5 h) of the tar fractions produced yields of reducing sugars in the range of 80–85% and an overall yield of more than 50% glucose from cellulose. According to Shafizadeh, the type of atmosphere during pyrolysis affects the pyrolytic reactions. In the presence of oxygen pyrolytic depolymerization, oxidation, and dehydration are accelerated. Depolymerization occurs more slowly in the presence of an inert atmosphere, but formations of unwanted by-products of oxidation and dehydration are also retarded. Moreover, the addition of a zinc chloride catalyst has been reported to cause decomposition of pure cellulose at a much lower temperature.

Fan et al. [67] observed a negligible change in the crystallinity index and surface area of Solka Floc upon pyrolysis pretreatment at 170 °C in air or helium atmosphere (Table 3.6); however, a marked increase in the hydrolysis rate was obtained for Solka Floc treated in helium atmosphere. They have proposed that this increase is due to depolymerization of cellulose.

3.3.1.3 High Energy Radiation

Digestibilities of cellulose and lignocellulosics are enhanced by high energy radiation [23, 99, 101, 103, 176, 177, 181, 243, 249]. The radiation of pure cellulose results in oxidative degradation of the molecules, dehydrogenation, destruction of anhydroglucose units to yield carbon dioxide and cellulosic chain cleavage [16, 82, 115, 123, 142, 146, 157, 181]. Fan et al. [68] showed that gamma irradiation of Solka Floc was very effective in increasing specific surface area, but not effective in decreasing the crystallinity index, as shown in Table 3.6. The increase in the surface area was due to extensive depolymerization. They observed that the hydrolysis rate increased only after the dosage exceeded a certain level.

According to Lawton et al. [146], who have studied the effect of radiation on lignocellulosics, the increased presence of phenolic groups in irradiated wood fibers indicates that lignin is affected by the treatment. Furthermore, pentoses and hexoses are generated as depolymerization products. Kumakura and Kaetsu [141] found that rise straw, chaff (rice husks), and sawdust (Japan cedar) had maximum digestibility when irradiated with 5×10^8 rad. Pritchard et al. [229] noted that the optimum digestibility of gamma irradiated wheat straw was 2.5×10^8 rad. Irradiation appears to be strongly species selective; for example, the digestion of aspen carbohydrate is essentially complete after an electron dosage of 10^8 rad, while spruce is only 14 % digestible at this dosage [50].

The digestibility can be increased by milling the substrate before irradiation [229], adding nitrate salts prior to irradiation [51], or irradiating in the presence of oxygen [28]. Blouin and Arthur [28] have stated that the total dosage and not the dosage rate appears to have the greatest effect on cellulose. According to Millett et al. [180], the technique lacks commercial appeal based on an estimated overall cost. Another effective radiative pretreatment is nitrite photochemical radiation. Here cellulose is irradiated by ultraviolet light in an aqueous solution of sodium nitrite [50].

3.3.1.4 High Pressure Steaming

High pressure steaming is a pretreatment method in which the substrate to be pretreated is steamed under pressure, at high temperatures and generally without

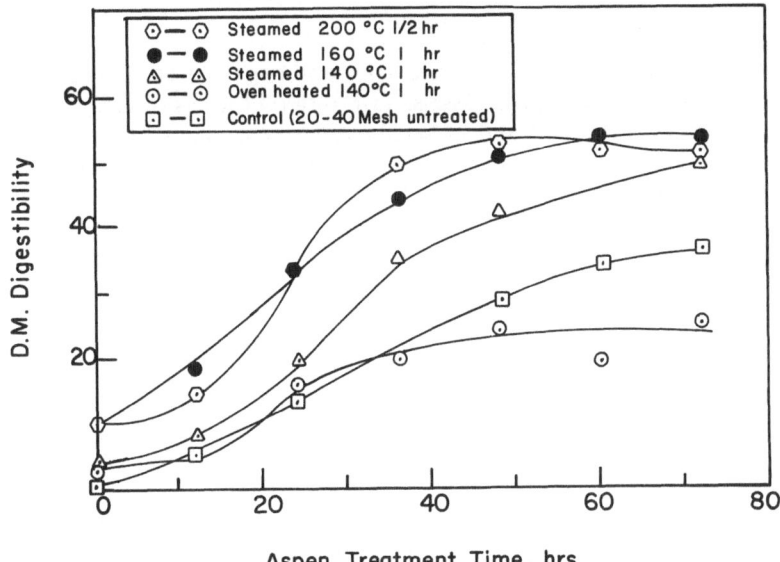

Fig. 3.10. Effect of steaming conditions on the dry mass digestibility of ground aspen chips [24]

the addition of chemicals [161, 200, 268]. The process cleaves the acetyl groups and provides an acidic medium conducive to hydrolytic action.

Bender et al. [24] reported on the *in vitro* digestibility of various hardwoods and softwoods pretreated with high pressure steam. The results for aspen chips are given in Fig. 3.10. When aspen was ground after steaming, its digestibility was 56.3 %; steamed unground aspen had a digestibility of 50.8 %, compared to the digestibility of one-week ball-milled aspen of 46.1 %. Heaney and Bender [107] pretreated aspen chips for about two hours at 100–115 psi (160–170 °C); this yielded an acceptable product for sheep for up to 60 % of the total ration. The sheep maintained normal weight gains.

Nesse et al. [186] autoclaved manure fibers for 5, 10, and 30 min at temperatures ranging from 130 to 200 °C and found the optimum treatment to be steam treatment for 30 min at 175 °C.

3.3.1.5 Extrusion and Expansion

Pretreatments similar to high pressure steaming are moist-heat expansion (extrusion) and dry-heat expansion (popping), both of which enhance feed efficiency of grains in animal feedlots [297, 320]. Data reported by Han and Callihan [97] show that extrusion pretreatment is ineffective in increasing the digestibilities of rice straw and sugarcane bagasse. The substrates, however, were pretreated only for a maximum of 90 s.

Brenner et al. [32] pretreated newspaper by means of the extrusion technique for acid hydrolysis and obtained remarkable results. They have suggested that extrusion may be a promising pretreatment method for acid hydrolysis.

3.3.1.6 Microwave Treatment

In a study by Ooshima [213], rice straw and bagasse were subjected to microwave irradiation in sealed glass vessels. Two different levels of moisture contents were considered, namely, 84 and 94%. The rate of enzymatic hydrolysis was enhanced markedly by this pretreatment, e.g., the rate for pretreated bagasse was 3.2 times that of untreated bagasse.

3.3.2 Chemical Pretreatments

Chemical pretreatments have been used extensively for the removal of lignin surrounding cellulose and for destroying its crystalline structure. Traditionally, the paper industry has been pulping cellulosic materials for delignification to produce high strength, long fiber paper products [37]. However, these processes are deemed to be overly severe and expensive for pretreatment of lignocellulosics. The possibility of developing effective and less expensive pretreatment methods by modifying the existing pulping processes deserves consideration [252].

3.3.2.1 Alkalis

Sodium hydroxide. Up to the present time, pretreatment with caustic soda has been applied mainly towards enhancement of the digestibility of lignocellulosic materials for ruminants rather than pretreatment for hydrolysis [9, 10, 17, 19, 22, 39, 45, 49, 70–72, 83, 85, 86, 100, 102, 115, 134, 160, 173, 174, 184, 186, 215, 216, 253, 268, 284, 287, 292, 322, 332]. Dilute NaOH treatment of lignocellulosic material causes swelling, leading to an increase in internal surface area, decrease in the degree of polymerization, decrease in crystallinity, separation of structural linkages between lignin and carbohydrates, and disruption of the lignin structure.

Various substrates respond differently to sodium hydroxide treatment. Feist et al. [70] observed that the digestibility of softwoods with high lignin content increased slightly with NaOH pretreatment, while the digestibility of some hardwoods and other lignocellulosic materials, with low lignin content, increased significantly upon NaOH treatment.

Although NaOH pretreatment markedly affects the digestibility of hardwood and straw, it has little effect on the digestibility of cotton. Data presented by Moore et al. [183] show that the percent of reducing sugar as glucose is essentially unchanged upon NaOH pretreatment of cotton. The optimum amount of NaOH required to enhance the digestibility of aspen wood is 5–6 g of NaOH per 100 g of wheat straw, as seen in Fig. 3.11 [50, 321].

Ammonia. This pretreatment to increase the digestibility of straw was first patented by Lehmann in 1905 [154]. More recently, several investigators [97, 227, 276, 296, 315] employed liquid or gaseous ammonia as a strong swelling agent for cellulose; they observed a change in the crystalline structure from cellulose I to cellulose III [179]. Moore et al. [183] treated aspen with liquid ammonia and found that the percentage yield of reducing sugar as glucose increased from 11% for untreated

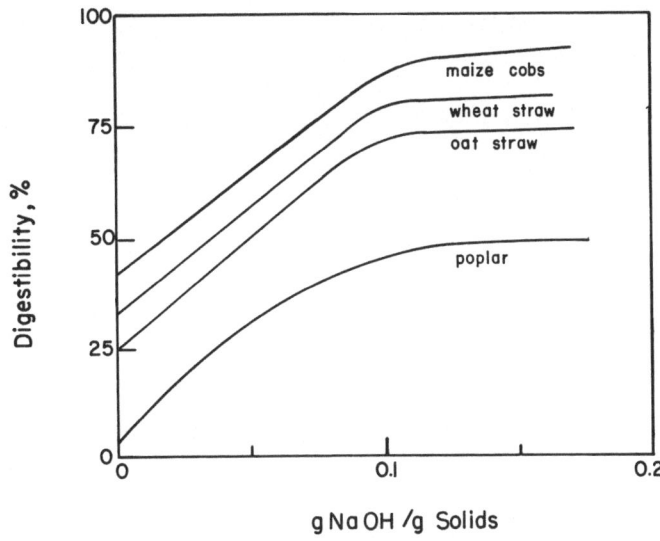

Fig. 3.11. Optimum amount of NaOH required to enhance the digestibility of various lignocellulosics [51]

aspen to approximately 36% for NH_3 treated aspen. Waiss et al. [296] achieved a 10% increase in the enzymatic digestibility after pretreatment with 5% NH_3 at room temperature for 30 days. Moore et al. [183] also pretreated cotton with liquid ammonia, and achieved a substantial increase in the percentage yield of reducing sugars. This increase was attributed mostly to the swelling in water subsequent to the pretreatment. Pretreatment with ammonia has also been used to enhance the digestibility of lignocellulosics for ruminants.

Ammonium Sulfite. Ammonium sulfite has been employed mainly in a conventional pulping process; the process has been modified by Clarke and Dyer [40] to increase the digestibility of lignocellulosics for animal feed. In this process, pulp is made by treating shredded Douglas Fir with $(NH_4)_2SO_3$ under high pressure and elevated temperature. The resultant pulp has a residual lignin content of 15% and a dietary energy equivalent of medium quality hay when fed to steers at up to 70% of their total ration.

3.3.2.2 Acids

Acids serve primarily as catalysts for hydrolysis of cellulose rather than as reagents for pretreatment [32, 89, 105, 116, 137, 151, 160, 184]. Acid hydrolysis of cellulose on an industrial scale was started during the First World War, and it continues to be practiced extensively in the USSR [263].

Acid pretreatments prior to enzymatic hydrolysis have been studied for sometime. Acids proposed for pretreatments are sulfuric acid [313], hydrochloric acid [97], and phosphoric acid [265, 298].

61

Sulfuric Acid. Han and Callihan [97] pretreated sugarcane bagasse and rice straw with sulfuric acid. Sugar production from sugarcane bagasse was maximized by treating it with 50% H_2SO_4 at 121 °C for 15 min, followed by dilution of the acid to 1% and heating for 15 min at 121 °C; the sugar yield was 23%. Rice straw was treated with 3% H_2SO_4 at 130 °C for 4 h; however, little improvement in growth of *cellulomonas sp.* was observed.

Tsao et al. [290] also pretreated lignocellulosics with sulfuric acid. The resultant pentose fraction was separated by a mild acid; the remaining residue was dried and mixed with concentrated acid (70–80%) to dissolve the cellulose, which eventually was precipitated from the solution by the addition of methanol. The precipitated cellulose was amorphous and could be easily saccharified by acid or enzymatic hydrolysis. This method was further investigated by Wilke [315], who obtained 35–49% overall sugar yield from corn stover.

Another sulfuric acid pretreatment was developed by Sasaki et al. [251], who employed concentrated acid. In their method, cellulose was dissolved in a short period of time; this was followed by precipitation of the dissolved cellulose with acetone. This process was applied to rice hulls, and 95% of the cellulose was converted to glucose by a commercial enzyme system in 24 h.

Knappert et al. [136] studied partial hydrolysis as a pretreatment for enhancing enzymatic hydrolysis of oak, corn stover, newsprint, and Solka Floc. They employed a continuous flow reactor, with temperature varying from 160° to 220 °C, and acid concentration from 0 to 1.2% at a fixed treatment time of 0.22 min. For all substrates, except Solka Floc, glucose yields were increased. In several cases, pretreatment resulted in 100% conversion of the potential glucose content of the substrate after 24 h of enzymatic hydrolysis. Investigations were carried out by Dunning and Lathrop [52] to establish an integrated process for utilizing agricultural residues. Numerous agricultural residues were pretreated by varying the acid concentration, temperature, and time. Pretreatment with 4.4% acid, at 100 °C for 55 min was found to give significantly improved results upon acid hydrolysis.

Hydrochloric Acid. Rice straw was pretreated with 3% HCl at 130 °C for 4 h by Han and Callihan [97]. Compared to the control, the cell yield of *Cellulomonas sp.* grown on the treated substrate decreased.

Phosphoric Acid. Walseth [298] investigated the possibility of pretreating cellulose with phosphoric acid. The pretreatment was carried out using 85% phosphoric acid and storing the samples at 2 °C for two different periods of time, namely, 10 min and 2 h. The substrates thus obtained were swollen and amorphous and hence, were highly reactive to enzymatic degradation. Approximately 80% digestion was attained in 100 h under appropriate conditions of hydrolysis. Ghose and Kostick [79] have discussed the possibility of using 85% phosphoric acid to swell cellulose as a means of pretreatment.

Stone et al. [265] observed increased swelling with an increasing phosphoric acid concentration. Furthermore, they observed a linear relationship between the initial hydrolysis rate and accessible surface area of cellulose swollen by phosphoric acid.

3.3.2.3 Gases

Pretreatment with gases has the advantage in that it facilitates uniform penetration throughout the substrate, thereby giving uniform coverage. On the other hand, a gaseous medium is more difficult to handle than a liquid medium, and the recovery of the former for reuse poses more problems than that of the latter [315].

Chlorine Dioxide. Chlorine dioxide is an active agent of the chlorite pulping technique, which utilizes sodium chlorite in an acetic acid solution to solubilize lignin. Saarinen et al. [248] improved the digestibility of birch wood for ruminants by means of the chlorite technique.

By passing ClO_2 through a bed of dried and ground wheat straw, Sullivan and Hershberger [272] determined that the maximum increase in the in vitro digestibility of treated wheat straw was 43% more than that of untreated wheat straw. They postulated that the increase was attributable to decomposition of lignin by chlorine and its oxides.

Nitrogen Oxides. A pulping scheme involving nitrogen oxides was investigated by Brink et al. [33, 34]. It gave a rapid rate of delignification and a higher overall sugar yield than conventional pulping processes at comparable lignin contents. These promising results indicate that nitrogen oxides may provide an effective pretreatment for enzymatic hydrolysis.

Nitrogen oxides as pretreatment agents for lignocellulosics was examined by Borrevik et al. [29] and Wilke [315]. Wilke pretreated 100 g of wheat straw in a 22 l vessel with 5 g of NO, followed by an addition of 6 g of O_2. The treatment lasted for 24 h at 25 °C. The best result obtained was a xylose yield of 69% based on the original xylan content of straw. By applying NO first to a suspension of cellulose and letting it stand for a few minutes before feeding oxygen, NO was able to permeate uniformly throughout the substrate prior to the reaction of NO with O_2 to form NO_2. Nitrogen dioxide then reacts with water to form nitric acid, which oxidizes cellulose and degrades lignin.

Sulfur Dioxide. Sulfur dioxide pretreatment is relatively unexplored as evident from the scarcity of publications. According to Dunlap et al. [51], sulfur dioxide pretreatment may be effective for lignocellulosics. In their process, gaseous SO_2 is reacted with moist lignocellulosics at 120° for 2–3 h. This treatment apparently disrupts the lignin-carbohydrate complex and depolymerizes the lignin. Millett et al. [180] studied the reactions of various species of wood with gaseous sulfur dioxide for a period of 2–3 h at room temperature and gas pressure of 30 psi without free water. Impressive results were obtained for both hardwoods and softwoods. Most of the hardwood carbohydrates were converted to sugars after being pretreated with SO_2 for 2 h, and 70–85% of the softwood carbohydrates were converted to sugars after being pretreated for 3 h.

Ozone. According to Schurz [252] ozone may be effective in degrading lignin without producing excessive amounts of pollutants. In an ozone pretreatment of wheat straw, Binder et al. obtained a drastic increase in the biodegradability of cellulose [65]. Ozone was observed to attack both lignin and carbohydrates, though the rate of reaction with the latter is slower. A 50% reduction in the original lignin content was deemed to be optimal. A similar and independent observation has also been made by Fan et al. [65].

3.3.2.4 Oxidizing Agents

Oxidizing agents for the pretreatment of lignocellulosics are listed in Table 3.7 [97]. These agents cause structural modification of cellulose by penetrating into the cellulose and then oxidizing it. Han and Callihan [97] have noted that some oxidants penetrate and react with both the crystalline and amorphous regions of cellulose, while other oxidants only attack the amorphous regions. The oxidizing agents existing in gaseous form have already been considered under the gas pretreatment section.

Table 3.7. Oxidizing agents for cellulose [97]

Sodium chlorite ($NaClO_2$)	Sodium hypochlorite (NaOCl)
Potassium bromate ($KBrO_3$)	Hydrogen peroxide (H_2O_2)
Potassium Iodate (KIO_3)	Nitrogen dioxide (NO_2)
Potassium permanganate ($KMnO_4$)	Chlorine dioxide (ClO_2)
Potassium peroxydisulfate ($K_2S_2O_8$)	Sulfur dioxide (SO_2)
Potassium perchlorate ($KClO_4$)	Ozone (O_3)

Hydrogen Peroxide. The action of H_2O_2 with Fe^{2+} as a catalyst causes cotton cellulose to oxidize and decompose into CO_2 [90, 138]. The reaction is thought to be similar to what takes place when brown-rot fungus degrades lignocellulosics. Elmund et al. [54] treated feedlot waste at a slurry concentration of 5% with 150 ml H_2O_2 and 10 ml ferrous sulfate at 22 °C for 6–8 h. The resultant substrate gave 46% cellulose conversion upon 24 h of enzymatic hydrolysis.

Peracetic Acid. To delignify corn stalks, sawdust from broad leaved trees, and sawdust from coniferous trees, Toyama and Ogawa and Toyama [283, 285, 287] utilized a 20% peracetic acid consisting of acetic anhydride and 35% hydrogen peroxide, 1 : 1 by volume. A significant increase in the rate of enzymatic hydrolysis was observed. Fan et al. [69] obtained a drastic increase in the digestibility of wheat straw upon peracetic acid pretreatment. This increase was attributed to the extensive delignification achieved by this pretreatment.

3.3.2.5 Cellulose Solvents

Cellulose solvents, such as cadoxen and CMCS, are able to swell and transform solid cellulose into a soluble state [228, 274]. This ability to dissolve cellulose has been exploited as a means of lignocellulosic pretreatment. Once cellulose is dissolved, the major factors that deter its degradation, high crystallinity and the

Table 3.8. Solvents for cellulose [291]

Calcium thiocyanate
Strontium thiocyanate
Quarternary pyridinium salts
Hydrazine hydrate
Benzyl trimethylammonium hydrate
Methylamine in DMSO
Triethylamine oxide
Cuprammonium hydroxide
bis (β-γ-dihydroxy propyl) disulfide
CMCS
Cadoxen
Phosphoric acid
Nitric acid
Hydrochloric acid
Sulfuric acid

---

presence of lignin, can be reduced. Turbak et al. [291] reported on a vast array of cellulose solvents, shown in Table 3.8. The cost of many of these solvents along with their toxicity may hinder their industrial application.

Cadoxen. Cadoxen is an alkaline solution, containing ethylene diamine and water. The solution can readily dissolve cellulose, which can be reprecipitated into a soft floc by adding excess water. Tsao and his colleagues [111, 144, 288–290] have shown that the dissolved cellulose can be hydrolyzed easily before the floc recrystallizes, that up to 10% of the cellulose can be dissolved in cadoxen at room temperature, and that approximately 90% glucose yield can be attained based on the total amount of cellulose reprecipitated from the solution. This pretreatment, however, has the disadvantages of solvent toxicity and high cost.

CMCS. CMCS solvent is composed of sodium tartrate, ferric chloride, sodium sulfite, and sodium hydroxide solution. The CMCS solvent dissolves cellulose which can be precipitated easily by adding excess water. Up to 4% cellulose can be dissolved in CMCS at room temperature. Tsao et al. [289] reported approximately 95% glucose yield based on the total amount of cellulose reprecipitated from the solution. CMCS dissolves less cellulose than cadoxen, but it is nontoxic.

3.3.2.6 Solvent Extraction

Extraction of lignin by means of organic solvents has been studied extensively in systems wherein mineral acids serve as catalysts and in uncatalyzed systems [104]. In both the catalyzed and uncatalyzed systems, organic acids liberated during the process have been shown to accelerate delignification. The main usage of solvent extraction has been for pulping. Kleinert [135] extracted lignin from spruce and poplar sawdust by a mixture of ethanol and water. Wilke [315] followed a similar approach; wheat straw was pretreated with a mixture of 50% ethanol-water at pH 6 at 180 °C and 230 °C for approximately 1 h. Overall reducing sugar yields of 39.6%

65

and 37.1 % were obtained at 180° and 230 °C, respectively, compared to a reducing sugar yield of 24% for untreated wheat straw. Wilke also studied the effect of a benzene-ethanol extraction upon 2 mm Wiley milled wheat straw and attained an overall reducing sugar yield of 38.2%.

In delignifying western cottonwood by an ethanol-water system and a variety of catalysts, Sarkanen et al. [250] have concluded that in the acid catalyzed systems, hardwoods and straws are readily delignified, providing pulps in generally higher yields than the conventional kraft or soda processes, whereas softwoods require high temperatures for delignification, and consequently, the pulp yields tend to be low. In base-catalyzed systems, ammonium and sodium sulfides are appropriate catalysts and the extraction method is applicable to a wide variety of species.

Selvam and Ghose [214] extracted lignin from rice husks with ethylene glycol. Their results have indicated that an optimum treatment involves heating of 8 g of rice husks with 100 ml of ethylene glycol at 170 °C in presence of an acid catalyst (conc. HCl) for 60 min; after 48 h of enzymatic hydrolysis, 46 mg of reducing sugar per ml was obtained. For rice husks, ethylene glycol pretreatment was observed to yield better results than milling, heating, or alkali pretreatment.

Organosolv delignification of southern yellow pine was investigated by April et al. [14] and Bowers and April [30]. Aqueous phenol was found to be more effective than aqueous n-butanol in an uncatalyzed system. Aqueous phenol led to a 90% delignification in 2 h at 205 °C. The yield of residual solids was 36%. Four hours of cooking with aqueous n-butanol at 205 °C resulted in a 40% delignification. Solvent recovery was 90–95% for n-butanol and 70–78% for phenol. April et al. [14] also investigated the role of aqueous n-butanol in extracting lignin from sweet gum. The maximum delignification (92%) was obtained at the reaction time of two hours at 200 °C. Linden et al. [156] extracted lignin and hemicellulose from wheat straw by geothermal water. A four-fold increase in the rate of enzymatic hydrolysis is reported.

Organosolv delignification as a pretreatment for enzymatic hydrolysis was investigated by Nolan et al. [197]. An integrated process scheme has been presented involving n-butanol pulping; preliminary experimental results have also been presented.

The economics of organosolv pulping was evaluated by Nguyen et al. [196]; they have concluded that in lignin extraction the recovery efficiency of ethanol is the most important step in determining the process economics.

3.3.2.7 Swelling Agents

Swelling agents, primarily strong electrolytic solvents, have been employed to pretreat cellulose. Two types of swelling are known; one is intercrystalline and the other intracrystalline. For example, water can penetrate and loosen only the amorphous region of cellulose; this is considered as an intercrystalline swelling agent. On the other hand, swelling agents, such as certain salts and alkali solutions (Table 3.9), affect both the amorphous and crystalline regions of cellulose; they are called intracrystalline swelling agents [252]. In other words, intracrystalline swelling agents are effective in loosening the crystalline region of cellulose. As an

Table 3.9. Swelling agents for cellulose [301]

Sodium hydroxide
Sulfuric acid
Nitrogen dioxide in dimethylsulfoxide
Zinc chloride
Ruthenium red
Phosphoric acid
Trimethylbenzylammonium hydroxide
Iron tartrate complex
Methacrylate embedding
Sodium zincate

---

illustration, the action of swelling agents untwists the outer skin on cotton fibers and causes it to split and form collars; the inner cellulose layers swell rapidly between the collars [301].

3.3.3 Biological Pretreatments

A biological pretreatment utilizes wood attacking microorganisms that can degrade lignin. The microorganisms can be classified into three categories, brown rots, white rots, and red rots. Brown rots mainly attack cellulose; white rots and red rots attack both lignin and cellulose. Present research is aimed towards finding those organisms which can degrade lignin. Table 3.10 lists the microorganisms that are under investigation [252]. Ander and Eriksson [8] classified four types of micro-organisms that degrade wood components; bacteria, soft-rot fungi, brown-rot fungi, and white-rot fungi. According to them, the white-rot fungi is the most promising for lignocellulosic pretreatment.

The white-rot fungi are basidiomycetes which have over 1,000 species; only about 25 species have been examined for lignin decomposition. Kirk and Harkin

Table 3.10. Some microorganisms that attack wood [252]

Brown rots (attack mainly cellulose)
  Piptoporus betulinus
  Laetiporus sulphureus
  Trametes quercina
  Fomitopsis pinicola
  Gloephyllum saepiarium

White rots (attack both lignin and cellulose)
  Fomes fomentarius
  Phellinus igniarius
  Ganoderma appalanatum
  Armillaria mellea
  Pleurotus ostreatus

Red rots (attacks both lignin and cellulose)
  Fomitopsis annosa

---

[133] resorted to a white-rot fungi to remove 42% of the lignin, 3% of the glucan (including cellulose), and 30% of the hemicellulose from birch wood. Degradation of lignin by white-rot fungi is a co-oxidative process [8]; consequently an accompanying carbon source is necessary (e.g., cellulose and/or hemicellulose). The fungi ligninases appear to attack the phenolic residues with demethylation and ring cleavage. Ericksson [56] and Eriksson and Goodell [58] achieved almost specific lignin degradation with cellulase-less mutants of white-rot fungi. By exposing *Polyporus adustus* to UV light, a regulatory gene, which controls the synthesis of cellulase, mannanase, and xylanase, was destroyed. Ander and Eriksson [8] reduced the lignin content of birch rods by 17% and stiffness by 10% by exposing them to a cellulase-less mutant white-rot fungi for six weeks.

In the work of Detroy et al. [48], lignin in wheat straw was degraded by *P. ostreatus*. Losses of lignin and cellulose were observed to be 22 and 14%, respectively, after 30 days, and 40 and 32%, respectively, after 70 days. The organism appeared to selectively degrade lignin during the first six days. A study with *Cyathus stercoreus* also resulted in preferential degradation of lignin; the experiment with pressure cooked wheat straw degraded 45% of the lignin, but consumed only 20% of the cellulose [2].

Biological delignification appears to be a promising technique, but its low rate has prevented its usage in large scale industrial processes. The possibility exists, however, that its rate can be accelerated through genetic modification of lignin degrading microorganisms or through partial physical and/or chemical processing of the substrate prior to biological pretreatment.

3.3.4 Economic Analysis of Pretreatments

Several effective pretreatments, both physical and chemical, have been identified by Fan et al. [69] for the enzymatic hydrolysis of wheat straw. Preliminary cost analyses were conducted, the results of which are enlisted in Tables 3.11 and 3.12. The cost of a chemical pretreatment included only the cost of chemicals, and that of a physical pretreatment only the cost of energy consumed by a laboratory scale

Table 3.11. Cost analysis of chemical pretreatment methods [69]

Type of pretreatment	Yield of sugar per kg of wheat straw, g/kg	Extent of hydrolysis after 8 h, g/l	Cost of chemical per kg of wheat straw, $/kg	Pretreatment cost based on sugar, $/kg
Caustic-AC	341.0	20.5	0.04	0.12
Sulfite-AC	252.9	16.0	0.32	1.26
Hypochlorite-AC	239.9	13.3	0.94	3.92
Peracetic acid	279.9	20.9	7.51	26.84
Butanol	72.4	4.7	2.86	39.49
Ethylene glycol-AC	241.2	18.2	11.25	46.53
Sulfuric acid	140.5	9.4	0.11	0.78
Standard	70.0	2.1	–	–

Table 3.12. Cost analysis of physical pretreatment methods [69]

Type of pretreatment	Yield of sugar per kg of wheat straw g/kg	Extent of hydrolysis after 8 h, g/l	Cost of energy consumed per kg of wheat straw $/kg	Pretreatment cost based on sugar, $/kg
Ball-milling, 8 h	255.98	9.1	1.48	5.82
Roller-milling				
¼ h	160.0	6.48	2.24	14.0
Fitz-milling,				
fine	100.0	3.0	0.01	0.1
Extrusion, with				
pressure	64.29	2.5	0.01	0.16
γ-Irradiation				
50 Mrad	207.02	5.6	0.1	0.48

apparatus. Therefore, the cost of physical pretreatments would be further reduced for an industrial scale operation.

For wheat straw the costs of chemical pretreatments varied from $0.04/kg for caustic pretreatment to $11.25/kg for ethylene glycol treatment. The costs of physical pretreatments varied from $0.01/kg for Fitz milling to $2.24/kg for rolling-milling.

Among the physical pretreatments, ball-milling gave the most promising results in terms of the hydrolysis rate and sugar yield. The pretreatment is clean and easy to operate, but the pretreatment time of 8 h may make it impractical in a large-scale operation.

Among the chemical pretreatments, caustic and sulfite pretreatments appear most promising. Caustic pretreatment was identified as a potential candidate for large-scale process development based on pretreatment cost, hydrolysis rate, and sugar yield. In addition, ethylene glycol pretreatment was notable because of its effectiveness and possibility of recovery through an appropriate scheme, which would significantly reduce its cost. The chemical pretreatments, however, have disadvantages which must not be ignored. These include use of specialized corrosion resistant equipment, need of extensive washing, and disposal of chemical wastes [69].

3.4 Kinetics of Enzymatic Hydrolysis of Cellulose

Enzyme kinetic studies have been concerned mainly with the homogeneous system, i.e., the action of a soluble enzyme on a soluble substrate. However, many important enzyme reactions occur in heterogeneous systems. Some of the heterogeneous enzyme reactions involve the action of insoluble enzymes on soluble substrates and the action of soluble enzymes on insoluble substrates. Certain mitochondrial enzymes, plant root surface phosphatase and immobilized enzymes are examples of the insoluble state enzymes which act on soluble substrates. Enzymes involved in the digestion of foodstuffs in animal intestine and enzymatic hydrolysis of cellulose and starch are examples of the action of the soluble enzymes on insoluble substrates. Considerable research on the action of insoluble enzymes on the soluble substrates has been performed with immobilized enzyme systems, whereas the action of soluble enzymes on insoluble substrates has not been studied extensively and remains as one of the least known areas of enzyme kinetics [151, 172].

Some of the kinetic studies of cellulases have been carried out by employing cellulose modifications, e.g., carboxymethyl cellulose (CMC) for an increased solubility [59]. The results obtained from the cellulose modifications cannot be applied directly to insoluble cellulose, because the physical properties of cellulose are altered significantly by the presence of substituent groups. Kinetic studies using soluble oligosaccharides provide only an approximate picture of the kinetics of the enzyme hydrolysis of a soluble cellulose [143, 155, 238, 280, 308].

Several reports dealing with the kinetics of cellulase on insoluble native cellulose have been published. Some of the models proposed in these reports are rather empirical [31, 153, 155, 294]. Several researchers [7, 35, 78, 108, 109] attempted to explain kinetic behavior of the insoluble cellulose-cellulase system by resorting to the Michaelis-Menten type kinetics with product inhibition effects. Kinetic models, each based on a combination of enzyme adsorption and Michaelis-Menten kinetics, have also been proposed [53, 112–114, 171]. Such models generally assume that the enzyme is adsorbed on the surface of the substrate and that the rate of digestion is proportional to the concentration of the adsorbed enzyme. A few kinetic models take into account the mode of action of each cellulase component on cellulose molecules with different degrees of polymerization [149, 212, 271]. These models are based on the Michaelis-Menten type kinetics for concurrent random and end-wise attack of the substrate involving end-product inhibitions and several types of enzymes.

A limited number of distributed parameter models, which consider both the mass transfer and reaction, have been proposed [245, 270]. However, experiments to test their validity were carried out with dextran-dextranase systems rather than with cellulose-cellulase systems. A feedback inhibition model of cellulose degradation by *Thermoactinomyces*, involving different enzyme fractions with an associated synergism, has been developed [15]. This model is similar to the microbial growth model.

The kinetics of enzymatic hydrolysis of insoluble cellulose depends primarily on three groups of factors: the structural features of cellulose, the nature of the enzyme system employed, and the mode of interaction between cellulose and enzyme. It appears that a kinetic model, which considers all these factors, has rarely been proposed. To derive a mechanistic kinetic model, the structural features of cellulose and the mode of action of the cellulase complex have been fully investigated by Lee et al. [153]. Furthermore, kinetic characteristics of the heterogeneous cellulose-cellulase system, such as mass transfer, adsorption and desorption of the enzyme, surface reaction and product inhibition, have been examined [153].

In this section, the kinetic characteristics of this heterogeneous enzyme reaction are described first and kinetic expressions for hydrolysis of insoluble cellulose by cellulase are then reviewed. In addition, the kinetics of cellulase on soluble cellooligosaccharides is discussed. Such information is essential in developing a kinetic model of enzymatic hydrolysis of native insoluble cellulose.

3.4.1 Kinetic Characteristics of the Heterogeneous Cellulose-Cellulase System

3.4.1.1 Mass Transfer Limitation – Diffusion of Enzyme

Formation of an enzyme-substrate complex is a prerequisite to the enzymatic hydrolysis of native cellulose. Since cellulose is an insoluble and structurally complex substrate, the formation of an enzyme-substrate complex can be achieved only by diffusion of the enzymes into the complex structural matrix of cellulose. Because of the heterogeneous nature of the process, the overall rate of reaction conceivably could be influenced by mass transfer resistances including the bulk phase resistance, resistance through the surface film around cellulose particles, and resistance through capillary pores of cellulose particles. The resistances, through the bulk phase and film adjacent to the particles, depend on the size of cellulose particles, cellulase concentration, and hydrodynamic conditions in the reactor such as the intensity of agitation and the Reynolds number corresponding to the flow through the reactor [153].

Van Dyke [294], experimenting to determine the magnitude of the bulk phase resistance, observed that agitation intensity had little effect on the hydrolysis, provided that the cellulose particles were completely suspended. The minimum rate of agitation (RPM) to suspend 8 wt-% cellulose was 100, which corresponds to the Reynolds number of 1,200. The Reynolds number, corresponding to a transition from the laminar to the turbulent flow pattern, was 1,000 for a three-bladed propeller used by Van Dyke in a baffled reactor. Other investigators [112, 125] obtained similar results in their experiments studying mass transfer limitations. When agitation speed exceeded 100 RPM, mass transfer resistances were negligible in a batch reactor containing pure cellulose as a substrate if its concentration was 2–10%. This indicates that the bulk and film resistance can be made virtually nonexistant under certain experimental conditions, including adequate mixing, proper particle size of cellulose, and proper enzyme concentration.

No experimental study has been reported on pore diffusion resistance inside cellulose particles. Kim [125] suggested that the macromolecular nature of enzyme molecules may cause pore diffusion resistance to be insignificant. In other words, the enzyme does not diffuse through the pores of cellulose particles because the size of an enzyme molecule is larger than most of the capillary pores of cellulose.

3.4.1.2 Adsorption and Desorption of Enzyme

Since adsorption of enzymes on the cellulose surface is a prerequisite for the hydrolysis of cellulose, it should be studied in detail if we wish to understand the kinetic behavior of the heterogeneous cellulose-cellulase system. One of the earliest works on adsorption of cellulase by cellulose was carried out by Halliwell [91]. He reported that the aqueous phase becomes relatively free of enzyme immediately after mixing cellulase and substrate because of adsorption of the enzyme by the cellulose; the extent of adsorption depends on the initial quantity of cellulose present and on subsequent incubation of the mixture.

Mandels et al. [166] also observed that cellulose strongly adsorbs cellulase under conditions optimum for enzyme action and that the extent of the enzyme or protein uptake is proportional to cellulose concentration. More than 90% of the enzyme and soluble protein were adsorbed from an unconcentrated cellulase solution by a mixture containing 10% cellulose, and about 90% of the enzyme, and more than 70% of protein were adsorbed from a concentrated cellulase solution by the same mixture. The actual uptake ranged from 0.005 to 0.064 mg protein per mg of cellulose. The maximum uptake of cellulase per volume of cellulose was attained when cellulose concentration was lowest. Adsorption increased as the average particle diameter decreased from 50 µm to 6.7 µm.

Adsorption of *Trichoderma* cellulase in aqueous solution on three kinds of cellulose was studied by Peitersen et al. [217]. They found that adsorption of protein and enzyme from the solution is largely independent of pH but is strongly dependent on the temperature and type of cellulose. The following Langmuir isotherm type equation was adopted to relate the adsorbed enzyme concentration to the initial free enzyme concentration:

$$E_{ads} = \frac{K_p E_{ads,m}}{1 + K_p E_0} E_0, \tag{3.10}$$

where

E_0 = protein concentration in the supernatant
E_{ads} = adsorbed protein
$E_{ads,m}$ = maximally adsorped protein
K_p = const.

Peitersen et al. [217] observed that generally an increase in the temperature reduced the maximum adsorbed protein. At the optimal hydrolysis temperature of 50 °C, the maximum adsorbed protein by Avicel, that by Solka Floc SW 40 and that by ball-milled Solka Floc were 0.027, 0.019, and 0.049 mg protein per mg cellulose, respectively. Peitersen et al. postulated that the variation among these values might stem from the differences in the total available surface areas among the various types of cellulose.

Others have observed differences in adsorbability among the cellulase components [314, 317]. Mandels et al. [166] reported that component (C_1), displaying an activity on filter paper, was more strongly adsorbed than component (C_x), displaying an activity on CMC. Wilke and Mitra [314] made a similar observation after determining adsorption equilibria of both C_1 and C_x activities at various concentrations of milled newsprint. The adsorption characteristics of cellulase components on bagasse, which contains appreciable amounts of cellulose and hemicellulose, were presented and discussed by Ghose and Bisaria [77] in the light of their role on the hydrolysis. According to them, simultaneous adsorption of

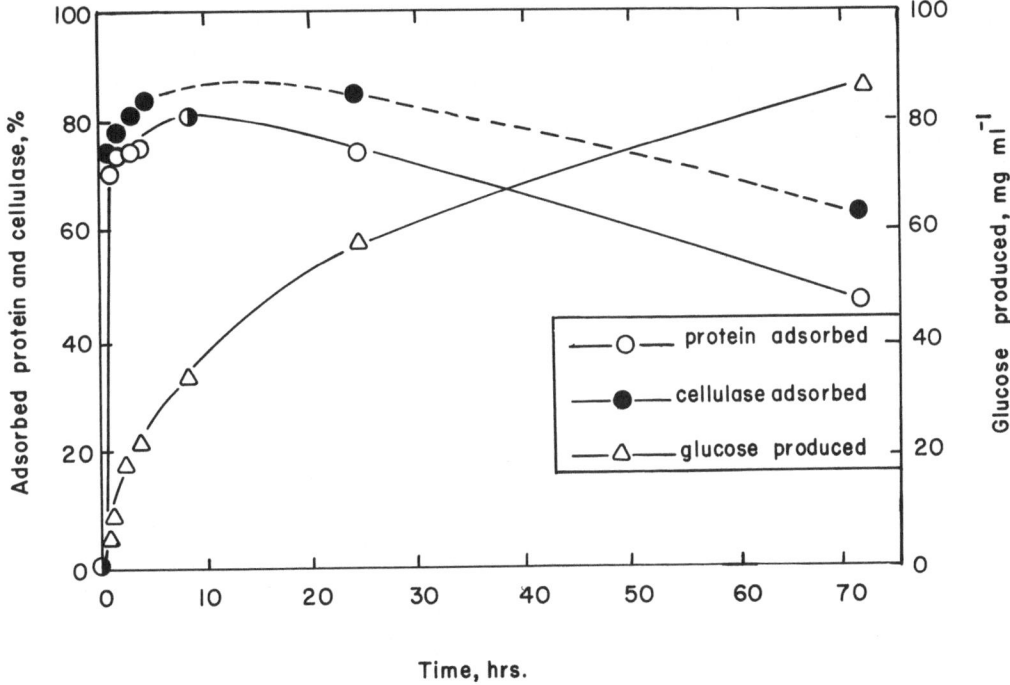

Fig. 3.12. Effect of time on adsorption of cellulase by cellulose [166]

exo- and endo-glucanase on hydrolyzable cellulose is the causative factor of hydrolysis that immediately follows the adsorption step; this supports the postulate of synergistic enzyme action.

The time course of adsorption of cellulase in a solution by cellulose, determined by Mandels et al. [166], is shown in Fig. 3.12. They mixed concentrated cellulase with 10% milled cellulose and incubated the resultant mixture in a shaker at 50 °C and at a pH of 4.4. Figure 3.12 shows that the initial uptake of cellulase was rapid with most of the enzyme removed from the solution after 20 min, but a slow uptake continued for several hours. Maximum adsorption was reached in approximately 8 h. Once the cellulose was supersaturated with cellulase, negligible uptake was observed; thereafter, the bulk concentration of the cellulase in the solution remained almost constant, and eventually increased slightly, indicating that the cellulase was released from the cellulose by the subsequent digestion of cellulose.

Other interesting experimental results on the behavior of adsorbed cellulase were obtained by Kim [125]; these results are shown in Fig. 3.13. The figure depicts typical digestion curves of a 6% suspension of ball-milled Solka Floc under various experimental conditions. Curve (A) shows the original batch reducing sugar production; curve (B) shows the reducing sugar production when the supernatant was removed and a buffer added. To obtain curve (B), the reaction mixture was centrifuged after 1 h of reaction time. Then, a fresh buffer solution was added to the precipitate, and the reaction was continued in a shaker incubator. As shown in Fig. 3.13, the reducing sugar production rate was nearly equal to that obtained in

73

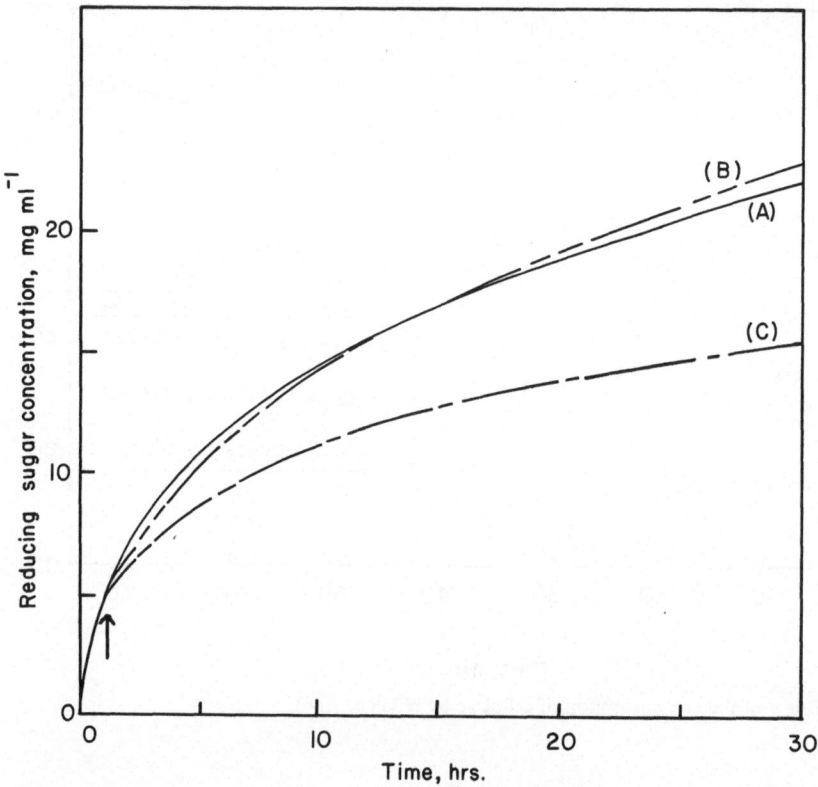

Fig. 3.13. Typical digestion curves of 6% ball-milled Solka Floc [125].
A Original batch, *B* supernatant removed; buffer added, *C* supernatant removed; buffer solution and 20 mg ml^{-1}, glucose solution added

the original batch (Curve A). This finding led Kim [125] to conclude that the enzyme adsorbed onto cellulose initially is primarily responsible for cellulose digestion. Nevertheless, his conclusion needs further verification because when the supernatant is removed, most of the enzyme may exist in the adsorbed state. Curve C shows the reducing sugar production when the supernatant was removed, and a buffer solution and a glucose solution with a concentration of 20 mg ml^{-1} were added. Curve C illustrates the effect of product inhibition on the cellulase action.

3.4.1.3 Surface Reaction, Fragmentation, and Changes of Surface Area During Hydrolysis

The reaction between cellulose and cellulase is heterogeneous, and thus the rate of reaction is expected to be proportional to the amount of cellulose surface accessible to the enzyme molecules. Cowling [42] and Cowling and Kirk [44] proposed the accessible cellulose surface to be the most important structural feature of cellulose for enzymatic hydrolysis. Stone et al. [265] reported that a linear relationship indeed exists between the initial rate of reaction and the initial surface area available

74

to the enzyme molecules. Phosphoric acid swollen cellulose was used as the substrate. Meanwhile, King [128] reported a different relationship between the rate of crystalline cellulose digestion and the available surface area. He determined the size distribution after each time increment and the corresponding amount of cellulose solubilized by multiplying the volume of each particle by the change in number of such particles per unit volume of solution. The surface area of particles in each size class was then calculated assuming that the particles were spherical. A linear relationship was found to exist between the rate of degradation of crystalline hydrocellulose particles and the square of their surface area, which suggests that the enzyme action is not exclusively a surface phenomenon.

According to King [127], the large cellulose fiber particles were fragmented into smaller ones (800–1500 Å) during the initial stage of reaction. These observations coincide with the early works of Halliwell [92] and Marsh [168], who reported that the early phase of enzymatic breakdown of cellulose is characterized by the formation of a large number of very short fragments, which increases to a maximum and then gradually decreases by conversion to soluble sugars. Therefore, in addition to the initial surface area of the cellulose, additional factors such as fragmentation of a cellulose particle and other structural features of cellulose should be considered in the kinetic study of the degradation of insoluble cellulose.

Fan et al. [66] measured the change in the specific surface area of Solka Floc during its hydrolysis by the filtrate of *Trichoderma* cellulase. The water swollen surface area of cellulose was measured by a solvent drying technique [175]. By applying the BET equation to nitrogen adsorption data, the specific surface areas during hydrolysis were determined as shown in Fig. 3.14. Total surface area was

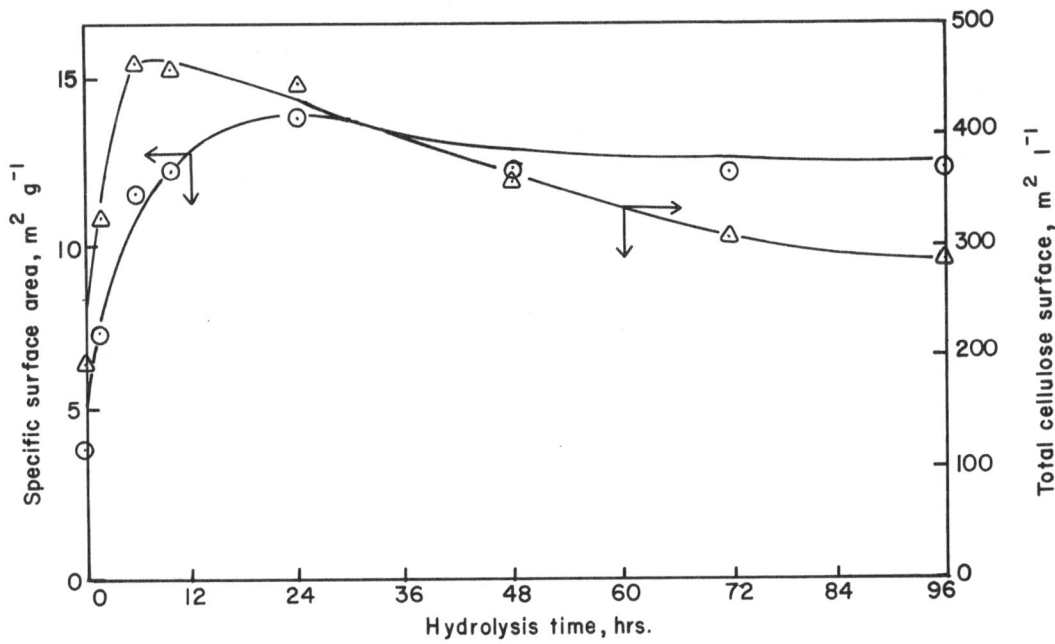

Fig. 3.14. The change of specific surface area and total surface area during hydrolysis [66]

evaluated by multiplying the residual cellulose weight per liter by the specific surface area. The specific surface area drastically increased from about $4\,m^2\,g^{-1}$ to $10.5\,m^2\,g^{-1}$ of cellulose during the first 6 h. Thereafter, it slowly increased until 24 h of hydrolysis, then the value leveled off to $12.2\,m^2\,g^{-1}$ of cellulose. The initial increase of specific surface area was observed to be mainly due to the fragmentation and particle size reduction. The reaction rate slowed even when a substantial amount of surface area existed. Consequently, Fan et al. [66] have concluded that the surface area may not be the main limiting factor in cellulose hydrolysis as presumed. To explain the slow-down of the reaction at the later phase, they have postulated that surface area is composed of two fractions, active and inactive, and that the rate of hydrolysis may be dependent on the active surface area rather than on the overall surface area.

3.4.1.4 Product Inhibition and its Mechanism

In enzyme reactions, the accumulated end products usually inhibit the rate of the forward reaction. A comprehensive review [162] has been published on the inhibition of cellulases and β-glucosidases caused by the physical factors, chemical actions, and natural inhibitors. Only the product inhibition is reviewed in this subsection.

Mandels and Reese [162] reported that cellobiose inhibited hydrolysis of cellulose with filtrates of the majority of the 36 organisms they tested. The inhibitive action of cellobiose was found to proceed competitively; in other words, cellobiose was competing with the substrate cellulose for active sites on the enzyme. Accordingly, the degree of inhibition depends on the relative affinities of the inhibitor and substrate for the active sites and on the ratio of the concentrations of the inhibitor and substrate. The inhibitory effect of products varies with the organism from which the cellulase is derived.

The product inhibition in the cellulose-cellulase system was observed to be large when the large soluble oligomer (cellopentaose) was present and to decrease rapidly with a decrease in the size of product molecules [190]. According to Ghose and Das [78], glucose mildly inhibits the hydrolysis of finely milled and heat-treated cellulose catalyzed by *Trichoderma viride* cellulases to an extent of only 40% at a concentration of 30%; whereas, cellobiose inhibits the reaction fairly severely (50%) even at a very low concentration. They analyzed their experimental results based on Henry's equation by assuming that the cellulase-cellulose system is an uncomplicated one-substrate reaction. They found that the plot of

$$\frac{1}{x}\log\left(\frac{a}{a-x}\right) \text{ against } \frac{t}{x}$$

did not yield the expected linearity for cellulosic substrate suspensions of 2%, 5%, and 10% in concentrated *T. viride* cellulase solution. In the plot, x is the concentration of cellulose, a is the initial cellulose concentration, and t is the reaction time. Thus, it was concluded that an inhibition mechanism is involved; furthermore from analyses of the Lineweaver-Burk plot the conclusion has been reached that the nature of inhibition can be represented by competitive product inhibition. The linear form of the product inhibition equation can be written as

$$\frac{1}{v} = \frac{K}{V}\left(1 + \frac{P}{K_p}\right)\frac{1}{(S)} + \frac{1}{V}, \tag{3.11}$$

where

   v = reaction rate
   V = maximum reaction rate
   P = product concentration
   K = Michaelis-Menten constant
   K_p = const
   (S) = substrate concentration.

Figure 3.15 shows the reciprocal or Lineweaver-Burk plots for no-inhibition, $10\,gl^{-1}$ glucose inhibition, and $5\,gl^{-1}$ cellobiose inhibition. These plots have yielded the following: (a) $K = 3.45\,mg\,ml^{-1}$, and $V = 1.3$ for the case of no inhibition, (b) $K_{p\text{-cellobiose}} = 28.90\,mg\,ml^{-1}$ for the case with $5\,mg\,ml^{-1}$ cellobiose inhibition, and (c) $K_{p,glucose} = 24.21\,mg\,ml^{-1}$ for the case of $10\,mg\,ml^{-1}$ glucose inhibition.

    Meanwhile, an enzyme inhibition mechanism, different from that of Ghose and Das [78] has been proposed by Stuck [269] and Howell and Stuck [109] for the hydrolysis of cellulose by the *Trichoderma viride* enzyme system. This mechanism postulates that the noncompetitive inhibition by cellobiose dominates the reaction kinetics. The model based on this noncompetitive mechanism has been reported to predict the progress of the reaction for an extended time more adequately than a competitive product inhibition model under various experimental conditions.

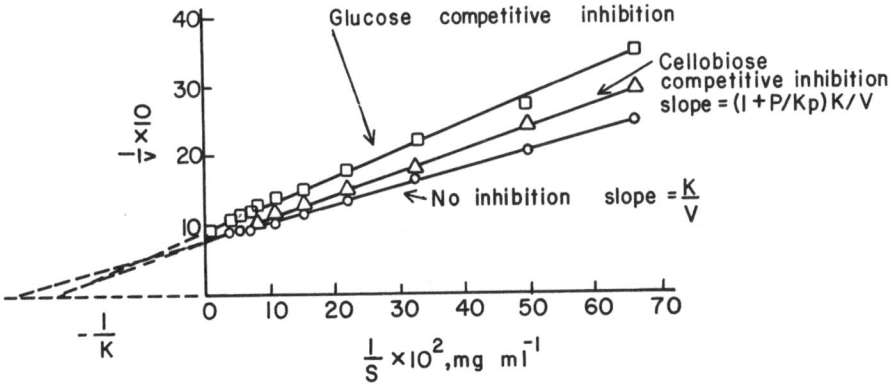

Fig. 3.15. Lineweaver – Burk plots for no inhibition, glucose and cellobiose inhibition in $<25\,\mu$ cellulose, enzyme activity 2.0 FP, Temp. 50 °C [78]

3.4.2 Kinetic Models of Insoluble Cellulose Hydrolysis by Cellulase

3.4.2.1 Empirical Representations of the Kinetics of Enzymatic Hydrolysis

An early empirical model was proposed by Karrer et al. [122], who used enzymes from the digestive juice of snails. These enzymes were found to hydrolyze regenerated cotton cellulose according to the so-called Schultz equation which will be elaborated later. Miyamoto and Nisizawa [182] investigated the kinetics of the hydrolysis of colloidal hydrocellulose catalyzed by cellulase preparations from several fungi. Analyzing their results according to the approach by Karrer et al. [121, 122], they have determined that the hydrolysis proceeds according to the following equation:
where

$$X = k\, t^m e^n, \qquad (3.12)$$

where
   X = extent of hydrolysis after time t
   k = empirical rate constant
   t = hydrolysis time
   e = concentration of enzyme
n, m = constants which are characteristics of each cellulase source.

Another empirical model, proposed by King [128], shows that the initial rate of hydrolysis of highly crystalline cellulose can be described by the Schultz law, which states that the extent of reaction is proportional to the square root of time, i.e.,

$$X = k\sqrt{t}, \qquad (3.13)$$

where
X = extent of hydrolysis
k = rate constant
t = time.

The investigation by Rautela and King [230] indicates that different types of crystalline cellulose yield distinct rate expressions. The rates of enzymatic dissolution of crystalline cellulose I and II obey the Schultz law, the rate of crystalline cellulose II/IV follows the second order kinetics, and the rate of crystalline cellulose IV follows no recognizable reaction order.

A more general description is given by Ghose [76] and Walseth [298] than by King [128]. They have theorized that the initial fast rate of hydrolysis may be expressed by the empirical expression

$$X = kt^n, \qquad (3.14)$$

where
X = extent of hydrolysis
k = rate constant
t = time
n = const.

According to Ghose [76] several factors, such as the heterogeneity of the substrate, interaction between the C_1 and C_x enzymes, and build-up of resistant cellulose during the course of hydrolysis, make it doubtful that the kinetic pattern, as suggested for the initial phase of reaction, is applicable to later stages of hydrolysis.

Kinetic expressions valid over the entire reaction period were proposed by Van Dyke [294] and Brandt et al. [31]. Van Dyke has resorted to the concept of multiple substrates within the matrix to formulate a model, which assumes that the formation of product is simply the summation of pseudo first-order reactions between the enzyme and substrate, that is,

$$\frac{dP}{dt} = \sum_{i=1}^{n} k_i C_i, \tag{3.15}$$

where
P = product concentration
C_i = concentration of cellulose component i
k_i = rate constant associated with component i.

To determine the concentration of the various components, the reaction progress curve was divided into regions, each of which was considered to have constant rate regions. The reaction curve in each of the divided regions was extrapolated to time zero. Thus, several intercepts which correspond to the different cellulose components could be obtained. Three cellulose components, insoluble amorphous cellulose, crystalline cellulose, and inaccessible cellulose, were reported to exist.

The model proposed by Brandt et al. [31] is similar to Van Dyke's [294], in that it also consists of a set of first order reactions to describe the process. Concentrations of different forms of cellulose were determined by location of inflection points.

The works of Van Dyke [294] and Brandt et al. [31] represent significant attempts to develop readily usable, kinetic expressions for the overall rate of cellulose hydrolysis. Nevertheless, their models do not take into account various enzyme-substrate reaction mechanisms, e.g., enzyme-substrate complex formation. Thus, their models should be considered as oversimplified semiempirical models.

3.4.2.2 Michaelis-Menten Mechanism Applied to an Insoluble Cellulose-Cellulase System

Amemura and Terui [7] attempted to apply the enzyme kinetics of the Michaelis-Menten type directly to an insoluble cellulose-cellulase system. The cellulolytic enzyme selected was obtained from *Penicillium variable*. This enzyme contains three cellulase components and degrades native cellulose directly into glucose without accumulating cellooligosides. Therefore, it is convenient for the kinetic study of insoluble substrates, as both the decrease in the weight of the substrate and increase in the reducing power in the incubation mixture can be readily determined as a measure of cellulase activity [6].

The effect of substrate concentration on the reaction rate, v, was investigated either by changing the initial substrate concentration $[S]_0$, or by following the

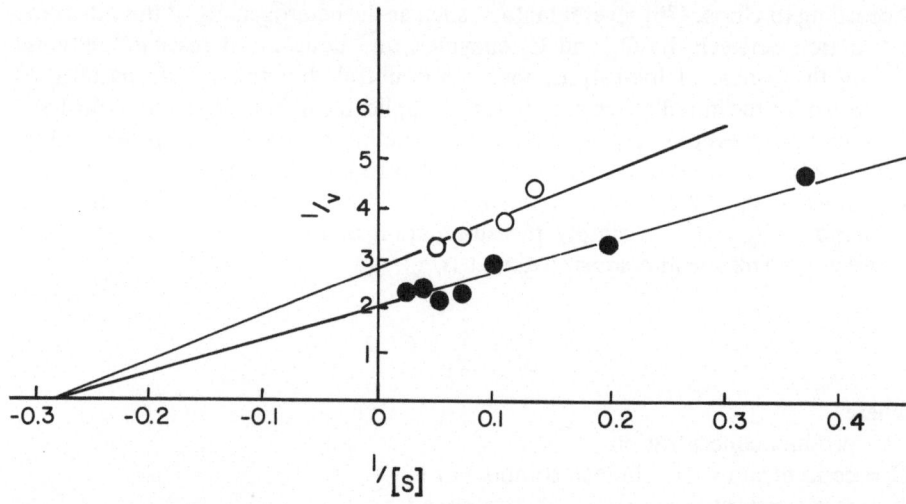

Fig. 3.16. Relationship between the substrate concentration and the initial rate of decomposition expressed by the Lineweaver-Burk plot [7]

residual substrate concentration, [S]. The data unexpectedly obeyed the classical Michaelis-Menten model as shown in Fig. 3.16. This figure plots $1/v$ against $1/[S]$ and the data are fitted to the integrated Michaelis-Menten equation. On the basis of this result, Amemura and Terui [7] have stated that neither the adsorption rate nor the penetration rate of enzyme molecules limits the overall rate of substrate dissolution and that the overall number of active sites in the substrate exposed to the enzyme attack remains nearly constant throughout the reaction. Note that their observations were based on the initial hydrolysis rate, where the structural changes can be negligible.

Since the cellulosic substrate usually used in hydrolysis contains 70% crystalline cellulose, which is difficult to be digested by cellulase, the rate equation based on the initial rate cannot be applied directly to the entire range of the reaction. Thus, the concept of the effective substrate concentration, $[S]_a$, has been introduced by Amemura and Terui [7]; it is defined as

$$[S]_a = \alpha [S]_t, \tag{3.16}$$

where

α = constant less than unity
$[S]_t$ = total substrate concentration.

Replacing $[S]_a$ in the Michaelis-Menton equation with $[S]_t$ gives rise to the following integrated rate expression.

$$t_r = \frac{1}{V'} \{[S_i](1-r) - K'_m \ln r\}, \tag{3.17}$$

where

$$V' = \frac{V}{\alpha}$$

$$K'_m = \frac{K_m}{\alpha}$$

$$r = \frac{[S]_{i+t_r}}{[S]_i}$$

$[S]_i$ = substrate concentration at time i
$[S]_{i+t_r}$ = substrate concentration at time $i + t_r$
t_r = time required for reducing the cellulose
concentration from $[S]_i$ to $[S]_{i+t_r}$.

A value of $3.75 \, gl^{-1}$ was obtained for the modified Michaelis-Menten constant, K'_m, from the initial rate experiments.

From the observation of the inhibition by the glucose produced, Amemura and Terui [7] introduced the product inhibition term in their kinetic model. Their analysis of the time course of reaction has indicated that the inhibition is of a competitive nature, and the inhibitor concentration, [I], is increased as the reaction proceeds due to the accumulation of glucose; the glucose concentration is proportional to $[S]_0 - [S]$, i.e., $[I] = c([S]_0 - [S])$. The following integrated form of the amended Michaelis-Menten equation was derived to fit the data.

$$t_r = \frac{1}{V'} \left\{ \left(1 - \frac{-cK'_m}{K'_i} \right) [S]_i (1 - r) - K'_m \left(1 + \frac{c[S]_0}{K_i} \right) \ln r \right\} \qquad (3.18)$$

Figure 3.17 shows the relationship between t_r and $[S]_i$, which yields

$$K'_i = 1.50, \ V' = 1.66 \, gl^{-1} h^{-1}, \ [S]_0 = 10.4 \, gl^{-1}$$
$$K'_m = 3.57 \, gl^{-1}, \ r = 0.8 \, .$$

From these results, Amemura and Terui [7] have concluded that the Michaelis-Menten equation is applicable even for the degradation of an insoluble substrate by the cellulase, if the total substrate concentration, $[S]_t$, is replaced by the effective substrate concentration, $[S]_a$, or if the accessibility of cellulase to the substrate is taken into account. Their work represents an early study on kinetics of enzymatic hydrolysis of cellulose, which attempted to apply the Michaelis-Menten kinetics with the product inhibition and substrate multiplicity. Although their model is based on the assumption of a constant α, no definitive experimental results have been obtained to determine the value of this constant.

The importance of product inhibition on kinetics of the enzyme system derived from *Trichoderma viride* was demonstrated by Ghose and Das [78], as indicated in a previous subsection. The initial hydrolysis of Solka Floc was determined in the presence and absence of glucose and cellobiose, and a significant product inhibition was observed.

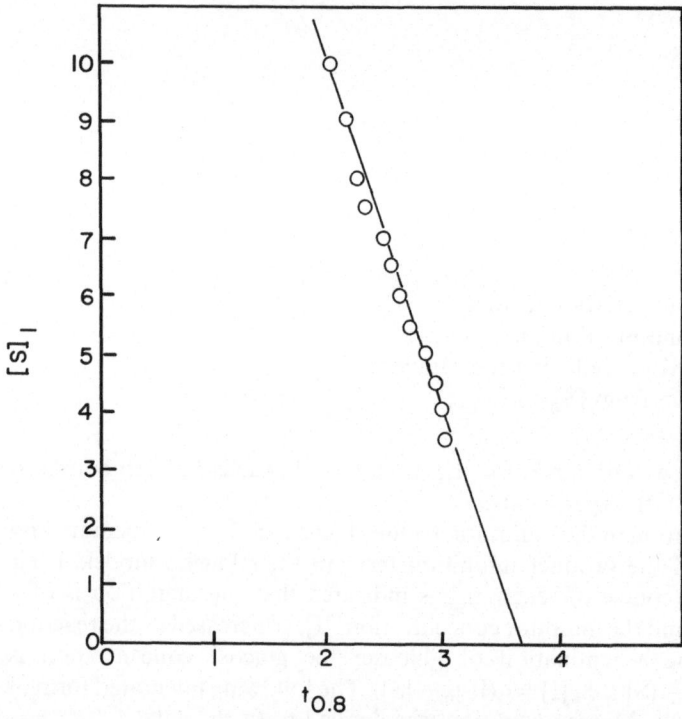

Fig. 3.17. Plot of Eq. (3.18) and experimental results showing the competitive nature of product (glucose) inhibition [7]

Meanwhile, Howell and Stuck [109] developed a non-competitive inhibition model to describe the hydrolysis of cellulose by the *Trichoderma viride* enzyme system. This enzyme was deficient in β-glucosidase, and thus the predominating end product over short to moderate time would be cellobiose. It was assumed that non-competitive inhibition by cellobiose dominates the reaction kinetics. The model is based on the mechanism,

$$[G_2] + E_1 \underset{k_{-1}}{\overset{k_{+1}}{\rightleftharpoons}} E_1[G_2], \tag{3.19}$$

$$E_1[G_2] \xrightarrow{k_2} E_1 + G_2, \tag{3.20}$$

$$E_1[G_2] + G_2 \underset{k_{-3}}{\overset{k_{+3}}{\rightleftharpoons}} E_1[G_2]G_2, \tag{3.21}$$

where

$[G_2]$ = cellobiose concentration in polymerized form
G_2 = cellobiose concentration
E_1 = enzyme concentration.

82

The application of the pseudo-steady state assumption to the enzyme complexes has yielded the following expression relating the substrate and product concentrations to time.

$$Vt = K_s \left(1 + \frac{S_0}{K_i}\right) \ln \frac{S_0}{S_0 - P} + \left(1 - \frac{K_s}{K_i}\right) P + \frac{P^2}{2K_i}, \tag{3.22}$$

where

S_0 = initial cellulose concentration, mol of anhydrocellobiose
$P = G_2$ = product (cellobiose) concentration
K_s = dissociation constant for the ES complex
K_i = dissociation constant for the EP complex
V = maximum reaction rate.

Kinetic parameters in this expression have been determined from the data obtained in batch experiments with durations of up to 15 h, by plotting the data on the Foster-Niemann coordinates. The resultant values of the parameters are: $K_i = 4.68 \times 10^{-3}$ mol cellobiose l^{-1}, $K_m = 9.56 \times 10^{-2}$ mol polyanhydrocellulose l^{-1}, $V = 2.92 \times 10^{-2}$ mol cellobiose $l^{-1} h^{-1}$ at an enzyme concentration equivalent to a filter paper activity of 1.5 units.

As a first approximation, Howell and Stuck [109] suggested the following equation;

$$t = 3.27 \left(1 + \frac{[G_2]_0}{4.68 \times 10^{-3}}\right) \ln \frac{[G_2]}{[G_2]_0 - G_2} - 6.58 \times 10^{-3} G_2 + \frac{[G_2]^2}{9.56 \times 10^{-3}}. \tag{3.23}$$

This equation depicts the progress of reaction with noncompetitive inhibition. The applicability of this model is limited to the initial short periods of time up to 10 h, corresponding to the extent of hydrolysis of 50–60%. After this period, the substrate multiplicity becomes significant in hydrolysis. Furthermore, the substrate concentrations in the experiments were lower than those usually employed in practice, namely, around 50 gl^{-1}.

By employing a similar product inhibition model and the Foster-Niemann plot, Brown and Waliuzzaman [35] investigated the kinetics of enzymatic hydrolysis of waste newspaper. The values of the parameters of the model obtained were almost identical to those for pure Solka Floc cellulose reported by Howell and Stuck [109]. It was suggested, therefore, that the data from the studies of pure cellulose would be directly applicable to the estimation of waste cellulose reactions.

Mangat [167] and Howell and Mangat [108] showed that the product inhibition kinetic model represents the experimental data well for the first 12 h of hydrolysis, but, thereafter, it overpredicts the rate of hydrolysis. To overcome this limitation, a model has been proposed by Howell and Mangat [108]. The main qualitative features of this model include: i) product inhibition by cellobiose is highly significant and controls the initial fast hydrolysis reaction, ii) a final portion of the substrate appears to be completely inaccessible to the enzyme, and iii) enzyme inactivation occurs according to a process other than product inhibition. Based on these premises, they proposed the following mechanism:

$$E + S \underset{k_{-1}}{\overset{k_1}{\rightleftharpoons}} ES, \tag{3.24}$$

$$ES \overset{k_2}{\longrightarrow} P, \tag{3.25}$$

$$E + P \underset{k_{-3}}{\overset{k_3}{\rightleftharpoons}} EP, \tag{3.26}$$

$$ES \overset{k_4}{\longrightarrow} D. \tag{3.27}$$

Equations (3.24)–(3.26) describe the competitive product inhibition mechanism, and Eq. (3.27) demonstrates that the E-S complex deactivates or denatures transforming into an inactivated enzyme-substrate complex (D). This inactivated complex plays no further part in the hydrolysis reaction. By applying a quasi steady-state assumption, the following integrated form of the equation was derived.

$$k_2 t = \frac{A}{\varepsilon} \ln \left[\frac{(E_0)}{(E_0) - \varepsilon(P)} \right] + \frac{B}{1+\varepsilon} \ln \left[\frac{(S_0)}{(S_0) - (1+\varepsilon)(P)} \right], \tag{3.28}$$

where

$$K_1 = \frac{k_{-1} + k_2 + k_4}{k_1}, \quad K_3 = \frac{k_{-3}}{k_3},$$

$$\varepsilon = \frac{k_4}{k_2},$$

$$A = \frac{K - \sigma \varepsilon}{(1+\varepsilon)(E_0) - \varepsilon(S_0)},$$

$$B = \frac{(E_0) \sigma (1+\varepsilon) - (S_0) K}{(E_0)(1+\varepsilon)(E_0) - \varepsilon(S_0)},$$

$$K = 1 + \varepsilon - \frac{K_1}{K_3},$$

$$\sigma = K_1 + (S_0)$$

The values of the parameters were estimated by a nonlinear least square minimization using numerical integration and optimization of the parameters. The values of the kinetic parameters obtained were $k_2 = 9.05$, $k_4 = 0.235$, $K_1 = 40.05$, and $K_3 = 2.40$. Howell and Mangat [108] used Eq. (3.28) to predict the progress of hydrolysis under the various experimental conditions and reported that the single enzyme model based on competitive product inhibition by cellobiose and a first order deactivation of the enzyme-substrate complex predicted the kinetic behavior of the hydrolysis reaction fairly well for an initial substrate concentration of 50 gl^{-1} and the time period up to around 90 h. The model appears to be applicable up to a duration of 3 or 4 days; in the latter period the substrate multiplicity strongly affects the kinetics.

84

3.4.2.3 McLaren's Approach to Enzyme Reactions in a Structurally Restricted System

A study was carried out by McLaren [171] to determine the enzyme kinetics of the hydrolysis of insoluble substrates by various types of hydrolytic enzymes. He has derived kinetic equations for such reaction systems based on the assumption that a soluble enzyme is adsorbed on the surface of a substrate in accordance with the Gyani-Freundlich isotherm and that the rate of digestion is proportional to the amount of the adsorbed enzyme. Thus,

$$E + A_s \rightleftarrows EA_s, \tag{3.29}$$

where
A_s = surface area of the substrate
E = enzyme concentration.

An additional assumption is that the solute is distributed between phase I (liquid) and phase II (solid) according to the Gyani-Freundlich equation, i.e.,

$$C_{II} = KC_I^n, \tag{3.30}$$

where
C_I = concentration in phase I
C_{II} = concentration in phase II
K = partition coefficient.

If phase II is a solid capable of only surface sorption, the surface concentration is given by

$$\frac{E_a}{A_s} = K \left(\frac{E_I}{V_I}\right)^{2/3} = KC^n, \tag{3.31}$$

where
E_a = number of moles of enzyme adsorbed on the surface, A_s, of phase II
E_I = number of moles of enzyme in the volume V_I of phase I at equilibrium
V_I = volume of phase I at equilibrium.

It is further assumed that

$$V_{initial} = k'(EA_s)$$

For a unit volume of a dilute suspension of the substrate with a unit surface area, this expression becomes

$$v_i = k' A_s K(E)^n / V_I = k'' K(E)^n \tag{3.32}$$

Since

$$(E_0) = (E) + (EA_s).$$

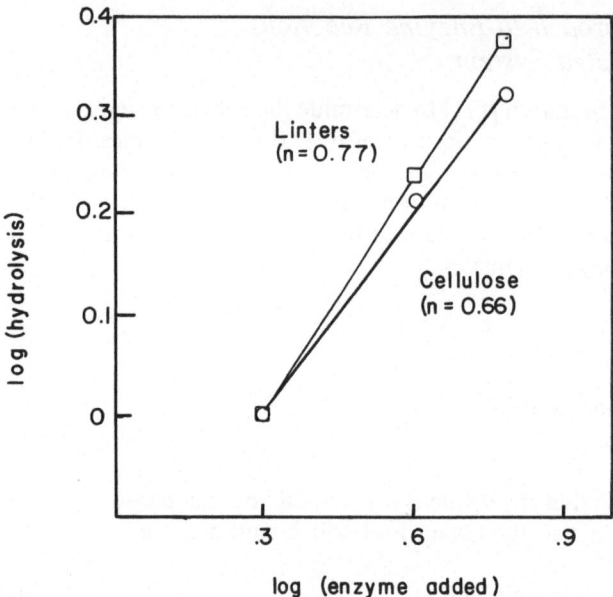

Fig. 3.18. Digestion of cellulose and swollen linters by cellulase [171]

Equation (3.32) becomes

$$v_i = k'' K [(E_0) - (EA_s)]^n. \qquad (3.33)$$

This shows that unlike the Michaelis-Menten kinetics, the initial rate of hydrolysis for an insoluble substrate system is not a linear function of the total enzyme in the system because n is less than unity. It is, however, proportional to the surface area of the substrate in contact with the enzyme. The equation has been successfully applied to the action of hydrolytic enzymes on starch granules, cellulose, protein gels, and oil emulsions. McLaren [171] has plotted the data obtained by Whitaker [305] for the cellulose hydrolysis as shown in Fig. 3.18. The equation is in good agreement with the data when n = 0.66 for cellulose and when n = 0.77 for the swollen linters. McLaren [171] has postulated that the difference in the value of n stems from the difference of penetrations through the substrates; a larger value of n for the swollen linter indicates that the enzyme penetrates farther into the swollen linters.

Note that in McLaren's theory for the rate of enzyme action, K and n are not arbitrary constants but are determined from the adsorption isotherms. Thus, the rate constant, K, may also be calculated from appropriate rate data. Strictly speaking, the Gyani equation is applicable only to a solute of small molecules. Therefore, we may use a theory of the adsorption of macromolecules of the protein type on surfaces, which is more sophisticated than the Gyani equation. In summary, the adsorptivity of enzymes on substrates is an important factor in kinetic analysis of cellulose hydrolysis. Unlike the Michaelis-Menten kinetics, the initial rate is not a linear function of the total enzyme concentration in the system. It is, however, proportional to the surface area of the substrate in contact with the solution.

McLaren [171] and McLaren and Packer [172] also derived an expression similar to Eq. (3.33), based on the assumption that the relation between E_0 and (E) is in the form of the Langmuir adsorption isotherm. McLaren [171] was one of the early investigators who attempted to describe this structurally restricted cellulose-cellulase system, which takes into account the adsorption of enzyme onto the surface of the substrate and the hydrolysis by the adsorbed enzyme. However, the applicability of McLaren's approach is limited only to very simple situations, because it ignores the substrate multiplicity, product inhibition, and mode of action of cellulase. The more advanced models, which consider the adsorption and Michaelis-Menten kinetics, have been developed subsequently and are described in the following section.

3.4.2.4 Kinetic Models Based on the Combination of the Enzyme Adsorption Equation and the Michaelis-Menten Type Equation

Huang [112–114] carried out a series of kinetic studies on insoluble cellulose-cellulase systems. His kinetic models combine an enzyme adsorption equation and an equation of the Michaelis-Menten type. He [112] first investigated the enzymatic hydrolysis of an insoluble amorphous cellulose (Walseth cellulose) by *Trichoderma viride* cellulase, which contained high levels of C_1 and C_x, and a low level of β-glucosidase. This system can be considered basically as a one-substrate, one-enzyme catalyzed reaction. Huang assumed that the enzyme is rapidly adsorbed onto solids on contact, and then gradually returns to the liquid phase as the reaction proceeds. Furthermore, end products, cellobiose and glucose were considered to be competitive inhibitors of cellulase action. Huang [112] depicted the mechanisms of the enzyme reaction as

$$E + S \underset{k_{-1}}{\overset{k_1}{\rightleftarrows}} X_1, \tag{3.34}$$

$$X_1 \overset{k_2}{\longrightarrow} E + P, \tag{3.35}$$

$$E + P \underset{k_{-3}}{\overset{k_3}{\rightleftarrows}} X_2. \tag{3.36}$$

He further assumed that the cellobiose to glucose ratio remains reasonably constant, the Langmuir adsorption model prevails, the second reaction is the rate determining step, steady state is attained, and the concentration of adsorbed enzyme, X_1, is much smaller than its saturated maximum value (X_{1m}). Based on these assumptions, he derived the following rate equation:

$$v = \frac{d(P)}{dt} = \frac{k_2 X_{1m} K_1 (E)_0 (S)}{1 + K_3 (P) + K_1 (E)_0 + K_1 (S)(X_{1m} - X_1)}, \qquad (3.37)$$

where

$$K_1 = \frac{k_1}{k_{-1}},$$

$$K_3 = \frac{k_3}{k_{-3}}.$$

Huang [112] substituted $(P) = (S)_0 - (S)$ into Eq. (3.37) and obtained the following expression upon integration of the resultant equation:

where

$$\frac{1}{\bar{v}} = \frac{1 + K_1 (E)_0 + K_3 (S)_0}{K_2 X_{1m} K_1 (E)_0} \frac{1}{(\bar{S})} + \frac{X_{1m} K_1 - K_3}{k_2 X_{1m} K_1 (E)_0}, \qquad (3.38)$$

where

$$\frac{1}{(\bar{S})} = \frac{1}{(P)} \ln \frac{(S)_0}{(S)_0 - (P)}$$

$$\bar{v} = \frac{(P)}{t}.$$

Huang [112] plotted the results of the hydrolysis experiments with amorphous cellulose according to Eq. (3.38), $1/\bar{v}$ against $1/(\bar{S})$. The experiments were carried out at several initial substrate concentrations and three initial enzyme concentrations. Table 3.13 shows the experimental and calculated values of the slopes and intercepts of the resultant plots for different values of $(S)_0$ and $(E)_0$. Note that at two enzyme levels, the data of $1/\bar{v}$ versus $1/(\bar{S})$ show sufficient linearity. Thus, the

Table 3.13. Experimental and calculated slopes and intercepts according to equation (3–38) [112]

$(E)_0$	$(S)_0$	Experimental		Calculated	
mg ml^{-1}		Slope	Intercept	Slope	Intercept
0.38	10.0	3.68	−0.33	3.69	−0.33
	7.2	2.68	−0.30	2.69	−0.33
	5.0	1.90	−0.32	1.90	−0.33
	3.6	1.38	−0.36	1.40	−0.33
0.25	10.7	6.00	−0.50	5.97	−0.50
	7.1	4.02	−0.51	4.01	−0.50
	5.2	2.98	−0.51	2.97	−0.50
	3.5	2.05	−0.54	2.05	−0.50
0.15	5.0	4.91	−0.88	4.75	−0.83
	3.8	3.68	−0.81	3.66	−0.83

data permit the evaluation of kinetic parameters k_2, K_1, X_{1m}, and subsequently K_3; they are $80.3\,h^{-1}$, $1.68\,ml\,mg^{-1}$, $0.305\,mg\,mg^{-1}$, and $5.61\,ml\,mg^{-1}$, respectively. Calculated data based on these parameters agree satisfactorily with the experimental data up to 70 % conversion.

Enzymatic degradation of usual cellulose such as Solka Floc would require a model consisting of two substrates, amorphous and crystalline cellulose. To apply the kinetic model for Solka Floc, Huang [113] transformed the kinetic expression of Eq. (3.38) into the following form.

$$\bar{v} = \frac{(P)}{t} = \frac{k_2 X_{1m} K_1 (E)_0}{X_{1m} K_1 - K_3} - \frac{1 + K_1 (E)_0 + K_3 (S)_0}{X_{1m} K_1 - K_3} \left\{ \frac{1}{t} \ln \frac{(S)_0}{(S)} \right\}. \quad (3.39)$$

He plotted the batch reaction data from the hydrolysis of Solka Floc, which contained both the amorphous and crystalline portions; specifically, $\dfrac{(P)}{t}$ was plotted against

$$\left\{ \frac{1}{t} \ln \frac{(S)_0}{(S)} \right\}.$$

Linearity was observed for all three enzyme levels at several solid suspensions and this linearity was interpreted as resulting from hydrolysis of the amorphous fraction of cellulose, whose concentration remains fairly constant throughout a batch run (up to 30 h of reaction time).

It is generally agreed that the crystalline cellulose is more resistant to degradation than the amorphous portion because hydrogen bonds are present in the former [42, 64]. Thus, a two-substrate system must be considered in kinetic modeling of the hydrolysis of cellulose, e.g., Solka Floc, where the amorphous and crystalline states co-exist. In addition, adsorption of cellulase onto solids and the effect of product inhibition should be included in analyzing this heterogeneous enzyme-catalyzed reaction. Huang [114] also constructed a kinetic model stipulating that enzyme E be adsorbed on crystalline cellulose S_c to form complex X_1, which, in turn, yields amorphous cellulose S_a. The enzyme is adsorbed on amorphous cellulose S_a to form complex X_2, which is degraded to soluble product P. The product then combines with the enzyme to form complex X_3, inhibiting the enzyme action. This mechanism can be depicted as

$$E + S_c \underset{k_{-1}}{\overset{k_1}{\rightleftarrows}} X_1 \underset{k_{-2}}{\overset{k_2}{\rightleftarrows}} E + S_a, \quad (3.40)$$

$$E + S_a \underset{k_{-3}}{\overset{k_3}{\rightleftarrows}} X_2 \overset{k_4}{\longrightarrow} E + P, \quad (3.41)$$

$$E + P \underset{k_{-5}}{\overset{k_5}{\rightleftarrows}} X_3. \quad (3.42)$$

Based on an assumption similar to that used in his earlier papers [112, 113], Huang [114] derived the following expression for the rate of reaction, v.

$$v = \frac{d(P)}{dt} = \frac{k_4 X_m K(E)_0 (S) f}{1 + K_5(P) + K(E)_0 + K X_m(S)}, \tag{3.43}$$

where

f = fraction of amorphous cellulose in the total cellulose
X_m = saturation amount of enzyme adsorbed per unit weight of cellulose

$$K = \frac{k_1}{k_{-1}} = \frac{k_3}{k_{-3}}$$

$$K_5 = \frac{k_5}{k_{-5}}.$$

By assuming that f is constant and that $(S) = (S)_0 - (P)$, integration of Eq. (3.43) gives

$$t = \frac{1 + K(E)_0 + K_5(S)_0}{k_4 X_m K(E)_0 f} \ln \frac{(S)_0}{(S)} + \frac{X_m K - K_5}{k_4 X_m K(E)_0 f}(P). \tag{3.44}$$

Upon rearrangement, the following linear form results

$$\frac{(P)}{t} = \frac{K_4 X_m K(E)_0 f}{X_m K - K_5} - \frac{1 + K(E)_0 + K_5(S)_0}{X_m K - K_5} \left\{ -\frac{1}{t} \ln \frac{(S)}{(S)_0} \right\}. \tag{3.45}$$

If $(E)_0$ is kept constant, plotting $(P)/t$ against

$$\left\{ -\frac{1}{t} \ln \frac{(S)}{(S)_0} \right\}$$

should yield a family of straight lines with $(S)_0$ as the parameter. The slope is a function of $(S)_0$ expressed in Eq. (3.45) as confirmed by Huang's experimental data [114]. Table 3.14 lists the slopes and intercepts. However, the intercepts at a given $(E)_0$ were not identical, which, according to Eq. (3.45), would result from different values of f for different values of $(S)_0$. In fact, it was observed that f decreases as $(S)_0$ increases. Huang [114] concluded that this is reasonable, because less enzyme becomes available for swelling per unit weight of crystalline cellulose as $(S)_0$ increases, and this would result in a small value of f. Therefore, he postulated that the negative intercepts result from the strong product inhibition.

An attempt was also made by Kim [125] to model and simulate cellulose hydrolysis with *T. viride cellulase*. The model takes into account such plausible mechanisms as bi-composition of cellulose, negligible mass transfer resistance, cellulase adsorption, formation and decomposition of the cellulose-cellulase complex, product inhibition, and cellulase deactivation. The assumptions were that the cellulose fiber is of a long cylindrical form, the concentration of cellulase adsorbed on cellulose is proportional to its concentration in the aqueous solution, and the adsorbed cellulase decays exponentially. The mechanism as suggested can be shown as follows:

Table 3.14. Slopes and intercepts of Solka Floc (BW-200) hydrolysis from a plot according to Eq. (3–45) [113, 114]

$(E)_0$	$(S)_0$	Slope	Intercept
$mg\,ml^{-1}$			
0.76	176	175	-0.115
	111	110	-0.137
	41.7	40.5	-0.171
	20.4	19.7	-0.167
	14.2	14.1	-0.180
0.44	111	111	-0.068
	41.7	41.7	-0.131
	20.4	20.8	-0.185
0.25	111	111	-0.019
	41.7	41.5	-0.046
	20.4	20.6	-0.085

Complex formation and decomposition:

$$C_A + E_A^A \underset{k_{-1}}{\overset{k_1}{\rightleftharpoons}} X_A \overset{k_2}{\longrightarrow} G_X + E_A^A, \tag{3.46}$$

$$C_C + E_A^C \underset{k_{-3}}{\overset{k_3}{\rightleftharpoons}} X_C \overset{k_4}{\longrightarrow} C_A + E_A^C. \tag{3.47}$$

Inhibition:

$$X_A + G_X \underset{k_{-5}}{\overset{k_5}{\rightleftharpoons}} Y_A, \tag{3.48}$$

$$X_C + G_X \underset{k_{-6}}{\overset{k_6}{\rightleftharpoons}} Y_C. \tag{3.49}$$

Enzyme deactivation:

$$E_A^A \overset{k_7}{\longrightarrow} E_A^d, \tag{3.50}$$

$$E_A^C \overset{k_7}{\longrightarrow} E_A^d, \tag{3.51}$$

where

C_A, C_C = concentrations of amorphous and crystalline cellulose, respectively
E_A^A = concentration of the enzyme active on amorphous cellulose
E_A^C = concentration of the enzyme active on crystalline cellulose
X_A, X_C = enzyme-cellulose complexes
G_X = products
Y_A, Y_C = enzyme-substrate-product complexes
E_A^d = deactivated form of the enzyme.

Table 3.15. Rate equations for hydrolysis of bi-composition cellulose [125]

Rate Equations

$$\dot{C}_A = -PA + \alpha X_c$$

$$\dot{C}_c = -PC$$

$$\dot{G}_x = \dot{\gamma} X_A - QA - QC$$

$$\dot{X}_A = \gamma X_A + PA - QA \qquad\qquad \dot{X}_c = \alpha X_c + PC - QC$$

$$\dot{Y}_A = QA$$

$$\dot{Y}_c = QC$$

where

$$PA = \frac{E_{A0}}{C_0^{1/2}} \frac{C^2 A}{C^{1/2}} \exp\left(-\varepsilon\tau\right) - X_A/P_{k1}$$

$$PC = \beta\left[\frac{E_{A0}}{C_0^{1/2}} \frac{C_c^2}{C^{1/2}} \exp\left(-\varepsilon\tau\right) - X_c/P_{k3}\right]$$

$$QA = \delta\left(X_A G_x^m - Y_A/P_{k5}\right)$$

$$QC = \eta\left(X_C G_x^m - \dot{Y}_c/P_{k6}\right)$$

$$\cdot \equiv \frac{d}{dt} \quad \tau = k_1 t \quad t = \text{time}$$

$$\alpha = \frac{k_4}{k_1} \quad \beta = \frac{k_3}{k_1} \quad \gamma = \frac{k_2}{k_1} \quad \delta = \frac{k_5}{k_1} \quad \eta = \frac{k_6}{k_1} \quad \varepsilon = \frac{k_7}{k_1}$$

The common type of product inhibition in an enzyme reaction results from the combination of products and free enzymes. Kim [125] postulated that this type of inhibition is not likely to affect cellulose hydrolysis since this system is heterogeneous, and thus, the enzyme adsorbed on the cellulose matrix is primarily responsible for the reaction. He also postulated that the most probable inhibitory mechanism results from a combination of the enzyme-substrate complexes and the products. Based on these postulations, a set of rate equations was derived as shown in Table 3.15. These equations were solved by using computer simulation subroutines such as CSMP, MIMIC, and GELG. Figure 3.19 compares the exact and approximate solutions, the initial rate (INI), and steady state kinetics (SS). Kim [125] pointed out that a simple Michaelis-Menten type kinetics does not adequately represent the system, and the steady-state solution usually obtained is invalid because of the complicated nature of the cellulose-cellulase system.

3.4.2.5 Kinetic Model Based on the Depolymerization Mechanism

Several kinetic models have been proposed that take into account the mode of action of different cellulase components on molecules with different degrees of polymerization. Suga et al. [271] considered the degradation of polysaccharide chains by endo-enzyme or exo-enzyme alone and also by various combinations of

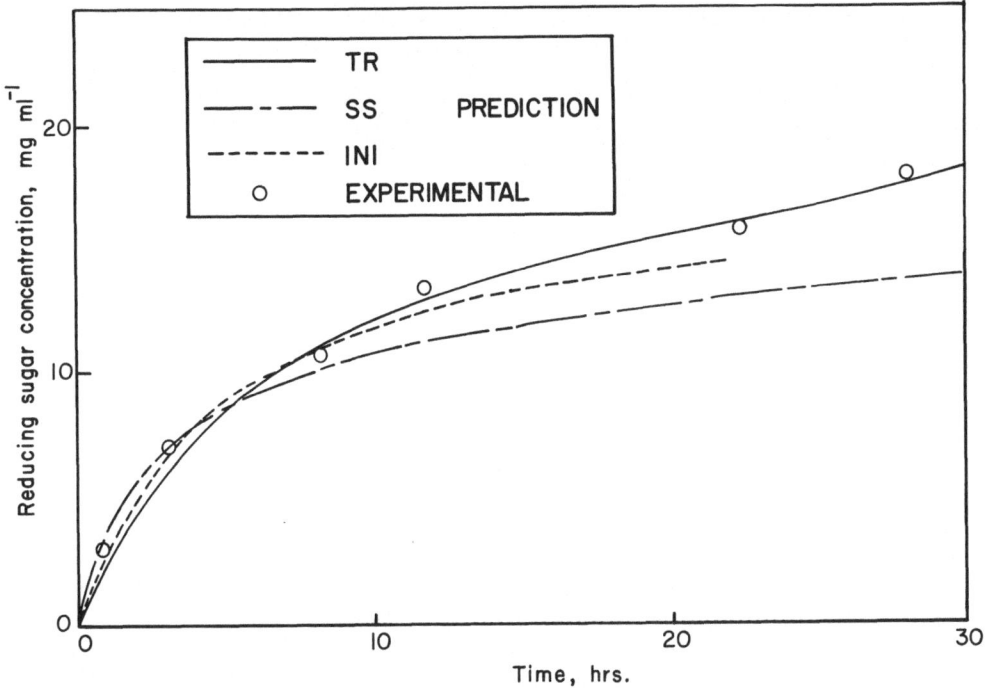

Fig. 3.19. Comparison between exact and approximate solutions, the initial rate (INI) and the steady state kinetics (SS) [125]

endo-enzyme and exo-enzyme. The reaction between the enzyme and substrate has been assumed to be of the common Michaelis-Menten type reaction, i.e.,

$$S + E \underset{k_{-1}}{\overset{k_1}{\rightleftharpoons}} ES \overset{k_3}{\longrightarrow} P + E. \tag{3.52}$$

By resorting to a quasi-steady state assumption, the following equation has been obtained for the rate of change in the concentration of a cellulose fragment with chain length i (C_i), in the presence of endo-enzyme alone.

$$\frac{dC_i}{dt} = - \frac{k_3 C_{Et}(i-1)C_i}{K_m + \sum\limits_{j=2}^{\infty} (j-1)C_j} + \frac{2K_3 C_{Et} \sum\limits_{j=i+1}^{\infty} C_j}{K_m + \sum\limits_{j=2}^{\infty} (j-1)C_j}, \tag{3.53}$$

where
C_{Et} = total enzyme concentration.

When an exo-enzyme, producing solely monomer (C_1) is present by itself, the rate of change of C_1 is given by

$$\frac{dC_1}{dt} = - \frac{k_3' C_{Et}' C_i}{K_m' + \sum\limits_{j=2}^{\infty} C_j} + \frac{k_3' C_{Et}' C_{i+1}}{K_m' + \sum\limits_{j=2}^{\infty} C_j}. \tag{3.54}$$

93

For the degradation of cellulose with various combinations of endo-enzyme and exo-enzyme, the following equation was obtained:

$$\frac{dC_i}{dt} = -\frac{k_3 C_{Et}(i-1)C_i}{K_m + \sum\limits_{j=2}^{\infty}(j-1)C_j} + 2\frac{k_3 C_{Et}\sum\limits_{j=i+1}^{\infty}C_j}{K_m + \sum\limits_{j=2}^{\infty}(j-1)C_j}$$

$$-\frac{k_3' C_{Et}' C_i}{K_m' + \sum\limits_{j=2}^{\infty}C_j} + \frac{k_3' C_{Et}' C_{i+1}}{K_m' + \sum\limits_{j=2}^{\infty}C_j}. \tag{3.55}$$

Equations (3.53) through (3.55) were solved to obtain the molecular weight distributions of cellulose fragments during the progress of hydrolysis. This model does not consider branch points of linkages, inhibition by products, and diffusion and adsorption of the enzyme through and on the substrate. However, the synergistic effect between endo-enzyme and exo-enzyme can be evaluated on the basis of this model.

While the model proposed by Suga et al. [271] considers only the endo-enzyme and exo-enzyme, the enzyme complex involved in cellulose hydrolysis contains endo-glucanase, exo-glucanase, cellobiosylhydrolase, and cellobiase [153]. The model has been extended by Lee et al. [148, 149] to the cellulase system with four enzyme components. The extended model includes the action of cellobiosyl-hydrolase and cellobiase; therefore, it describes more completely the action of the cellulase complex than the previous model [271].

Okazaki and Moo-Young [212] have proposed another extended model based on the Michaelis-Menten type kinetics for concurrent random and endwise attack of the substrate involving end-product inhibitions and three types of enzymes; endo-glucanase, cellobiosylhydrolase, and β-glucosidase. The model has served in investigation of the synergistic effect among the cellulase components, the dependency of the enzyme activity on the degree of polymerization of cellulose, and the substrate inhibition in cellulose degradation. Those interested in the mathematical details of the model may refer to the original paper [212].

The depolymerization models derived above are suitable for expressing the mode of action of each of the cellulase components and the synergistic action among them. In spite of this merit, each type of the depolymerization models contains a relatively large number of parameters. Moreover, these models ignore some significant factors in kinetic interactions such as the effect of the structural features of cellulose, mass transfer of enzyme and products, and the adsorption of enzyme.

3.4.2.6 Distributed Parameter Kinetic Models

A combination of the Fickian diffusion and Michaelis-Menten (M-M) kinetics was performed by Ross and Updegraff [245] to describe the rate of a diffussion-coupled biochemical reaction, specifically, the conversion of cellulose to protein by *Myrothecium vurrucaria*. In their model, Fick's second law is assumed to govern the diffusion, and the differential equation of the M-M rate mechanism is assumed to govern the reaction. Thus, the equation

$$\frac{\partial C}{\partial t} = D \frac{\partial^2 C}{\partial x^2} \qquad (3.56)$$

is obtained, which is governed by the following boundary conditions (B.C) and initial conditions (I.C.);

$$\text{B.C.;} \quad -D \frac{\partial C}{\partial x} = k \frac{C}{K + C} \quad \text{at} \quad x = 0, \qquad (3.57)$$

$$C = 0 \qquad \text{at} \quad x \to \infty, \qquad (3.58)$$

$$\text{I.C.;} \qquad C = 0 \qquad \text{at} \quad x = 0, \qquad (3.59)$$

where C is the concentration of the product. This nonlinear problem was solved by an approximate method (perturbation method) developed by Ross. When the principal interest is the product (protein) formation, the time integral of the product flux at the cellulose surface is obtained as

$$M = -\int_0^t D \frac{\partial C}{\partial x}\bigg|_{x=0} dt = kt - K \sqrt{\pi Dt}. \qquad (3.60)$$

This equation provides the basis for correlating experimental results. Ross and Updegraff [245] plotted the experimental data obtained by Updegraff [293] in two ways, namely, Mt^{-1} vs. $\sqrt{t^{-1}}$ and $M\sqrt{t^{-1}}$ vs. $\sqrt{t}$. These plots yielded two linear correlations, one with a slope of $-K\sqrt{\pi D}$ and intercept of k, and the other with a slope of k and an intercept of $-K\sqrt{\pi D}$. Such correlations permitted the identification of regions of diffusion control and reaction control. Since the model, as represented by Eqs. (3.56) and (3.60), is an oversimplification for all but the most elementary microbial processes, it can represent only very simplified kinetics.

Suga et al. [270] developed a model for an enzymatic reaction on an insoluble substrate. They investigated the kinetic behavior of the enzymatic breakdown of cross-linked dextran by a dextranase from *Penicillium funiculosum* as a model for enzymatic degradation of insoluble materials. This system is simpler than the cellulose-cellulase system, because only one enzyme is involved [304]. For a reaction between an insoluble substrate and an enzyme, the mass transfer of the enzyme through the cross-linked substrate should be considered. The reaction is assumed to be of the common Michaelis-Menten type [270].

$$S + E \underset{k_2}{\overset{k_1}{\rightleftarrows}} ES \overset{k_3}{\longrightarrow} P + E. \qquad (3.61)$$

Five other assumptions have been made; they include (i) diffusion of the enzyme and the released substrate fragments can be represented by Fick's first law, (ii) the diffusion coefficient of the soluble substrate fragments can be represented as a function of their average molecular weight, (iii) the average pore size of the particles is constant, (iv) the restricted diffusion coefficient of the enzyme and that of the released substrate fragments follow the equation proposed by Renkin [242], and (v) the Sephadex particles are spherical.

According to the model by Suga et al. [270], a material balance for the substrate fragments of chain length i over the differential volume of a spherical particle gives

$$\frac{\partial C_i}{\partial t} = \frac{1}{r^2} \frac{\partial}{\partial r} \left(D_s r^2 \frac{\partial C_i}{\partial r} \right) + R(C_i, C_E)$$

$$= D_s \frac{\partial^2 C_i}{\partial r^2} + D_s \frac{2}{r} \frac{\partial C_i}{\partial r} + \frac{\partial D_s}{\partial r} \frac{\partial C_i}{\partial r} + R(C_i, C_E), \qquad (3.62)$$

where
C_i = substrate fragments of chain length i
D_s = average restricted diffusion coefficient of soluble substrate fragments
C_E = concentration of the enzyme in particles.

By assuming random fission and constant values for k_3 and K_m for all chain lengths,

$$R(C_i, C_E) = - \frac{k_c C_E (i-1) C_i}{K_m + \sum\limits_{j=2}^{\infty} (j-1) C_j}$$

$$+ \frac{2 k_3 C_E \sum\limits_{j=i+1}^{\infty} C_j}{K_m + \sum\limits_{j=2}^{\infty} (j-1) C_j}. \qquad (3.63)$$

Combining Eqs. (3.62) and (3.63) yields

$$\frac{\partial C_s}{\partial t} = D_s \frac{\partial^2 C_s}{\partial r^2} + D_s \frac{2}{r} \frac{\partial C_s}{\partial r} + \frac{\partial D_s}{\partial r} \frac{\partial C_s}{\partial r} - \frac{k_3 C_E C_s}{K_m + C_s}, \qquad (3.64)$$

where C_s is the total sum of the concentrations of breakable substrates. The material balance for the enzyme over the differential volume of a spherical particle gives

$$\frac{\partial C_E}{\partial t} = \left(\frac{\partial^2 C_E}{\partial r^2} + \frac{2}{r} \frac{\partial C_E}{\partial r} \right) D_E. \qquad (3.65)$$

The restricted diffusion coefficient as proposed by Renkin [242] is

$$\phi = \frac{D_E}{D_{E0}} = \left(1 - \frac{a}{\gamma} \right)^2 \left\{ 1 - 2.014 \left(\frac{a}{\gamma} \right) + 2.09 \left(\frac{a}{\gamma} \right)^3 - 0.95 \left(\frac{a}{\gamma} \right)^5 \right\}, \qquad (3.66)$$

where
a = radius of enzyme
γ = pore radius.

96

Also, the restricted diffusion coefficient of released substrate fragments, D_s, is expressed as

$$D_s = D_{SF} \eta, \tag{3.67}$$

where

D_{SF} = average free diffusion coefficient of the fragments according to their average molecular weight, $cm^2 s^{-1}$.

Furthermore, the correction factor pertaining to diffusional restriction, η, is defined as

$$\eta = b\phi, \tag{3.68}$$

where b is an empirical constant. A material balance for the substrate in the solution gives

$$V_L \frac{dC_{sL}}{dt} = 4\pi R^2 n \overline{k}_s (C_{s,r=R} - C_{sL}) - V_L R(C_{sL}, C_{E_0}), \tag{3.69}$$

where

V_L = volume of the liquid phase
C_{sL} = concentration of the substrate in solution
R = particle radius
n = number of particles
$\overline{k}_S$ = overall mass transfer coefficient of the substrate.

Suga et al. [270] solved these equations numerically, subject to the appropriate boundary and initial conditions and determined the parameters of the model, such as K_m, V, K_d, D_{E_0}, D_{SF}, D_{ES}, a, and D_s experimentally. Their model was verified against experimental data from the degradation of the cross-linked dextran. Figure 3.20 compares the experimental data with the theoretical curves for the following values of parameters:

$K_m = 0.952 \times 10^{-6} \, mol \, cm^{-3}$, $V = 0.222 \times 10^{-8} \, mol \, cm^{-3}$,

$R \, (particle \, radius) = 0.01 \, cm$, $b = 0.009$, $a = 0.283 \times 10^{-6} \, cm$,

$\gamma = 0.560 \times 10^{-6} \, cm$, $c = 0.001 \, cm$.

The experimental data agree well with the theoretical curves. Thus, the model can be used to explain the behavior of the enzymatic breakdown of cross-linked dextran when the kinetics of the reaction and the physical properties of the substrate are known. This type of model may also be useful in interpreting the behavior of the enzymatic degradation of other insoluble substrates such as cellulose.

3.4.2.7 Mechanistic Kinetic Model

A comprehensive mechanistic kinetic model of cellulose hydrolysis is available [63]. It has been generated by combining models for several key aspects governing enzymatic hydrolysis of insoluble cellulose. The model assumes the following. (i)

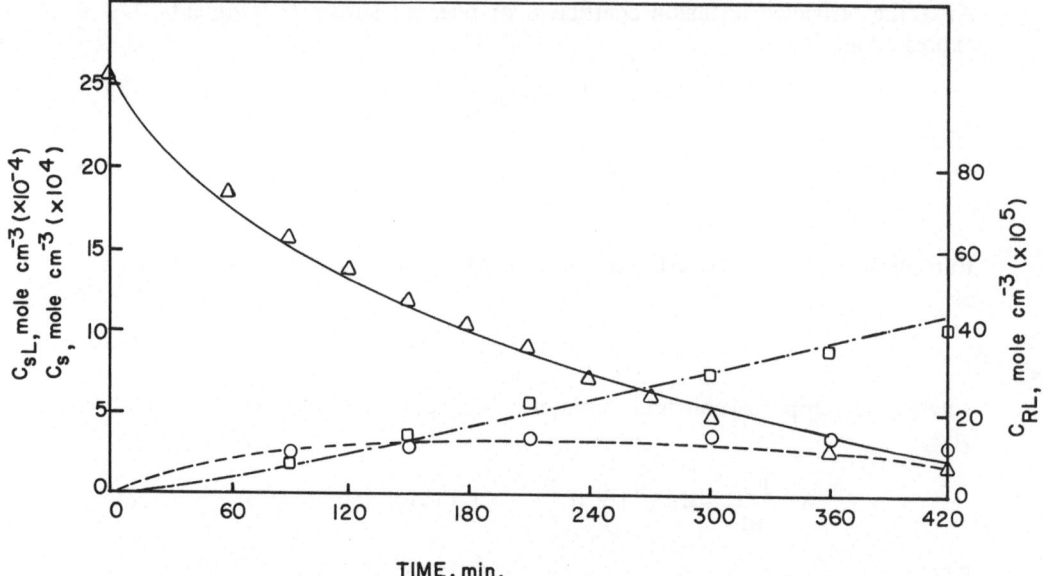

Fig. 3.20. Comparison between experimentally obtained change in substrate concentration in particle (△) and in solution (□), and concentration of reducing sugar (○) and theoretical curves [270]

The cellulose is hydrolyzed by the synergistic action of endo-β-1,4-glucanase and β-1,4-glucan cellobiohydrolase, and the resultant product, cellobiose is further hydrolyzed to glucose by the action of β-glucosidase. (ii) The two fractions of cellulose, crystalline and amorphous, are tightly interwoven so that they cannot be easily distinguished as two separate regions. Therefore, the bicompositional nature of cellulose is ignored. (iii) The overall hydrolysis rate is not influenced by the various mass transfer steps; thus, the heterogeneous system can be modeled like a homogeneous system. (iv) The hydrolysis reaction is controlled by the extent of soluble protein (enzyme) adsorption and the effectiveness of the adsorbed soluble protein to promote the hydrolysis. The extent of enzyme activity adsorption is approximately equal to the extent of soluble protein adsorption. (v) The hydrolysis of cellulose to cellobiose occurs at the surface of the cellulose by the adsorbed enzyme. This reaction is non-competitively inhibited by the soluble products, cellobiose and glucose. On the other hand, the decomposition of cellobiose to glucose occurs in the aqueous phase by unadsorbed β-glucosidase. The action of β-glucosidase is competitively inhibited by the product, glucose. (vi) The reduction in the hydrolysis rate with the progress of hydrolysis is influenced only by two factors, namely, the inactivation of adsorbed enzyme by the product and the transformation of cellulose into a less digestible form. (vii) The effective cellulose concentration can be represented by the average bulk concentration. Because it is difficult to express the substrate concentration in molar terms, unless such a concentration is expressed as the molar concentration of scissile bonds, the mass unit is utilized in describing the substrate and product concentrations.

98

The assumptions give rise to a mechanism for the enzymatic hydrolysis of cellulose involving two major steps, A and B; each step, in turn, is composed of several substeps as illustrated below.

A) Decomposition of cellulose to cellobiose (solid phase reaction) including adsorption of enzyme leading to the E-S complex formation,

$$(E) + (S) \rightleftarrows (ES) \qquad (3.70)$$

inactivation of the E-S complex by non-competitive product inhibition,

$$(ES) + (P_2) \rightleftarrows (ESP_2), \qquad (3.71)$$

$$(ES) + (P_1) \rightleftarrows (ESP_1), \qquad (3.72)$$

where the effective portion of the E-S complex is written as,

$$(ES)^e = (ES) - (ESP_2) - (ESP_1) \qquad (3.73)$$

transformation of the cellulose into a less digestible form,

$$(S) \rightarrow (S)^i \qquad (3.74)$$

and hydrolysis

$$(ES)^e \rightarrow (E) + (P_2). \qquad (3.75)$$

B) Decomposition of cellobiose to glucose (aqueous phase reaction) including E-S complex formation and hydrolysis

$$(P_2) + (E_\beta) \rightleftarrows (P_2 E_\beta) \rightarrow E_\beta + (P_1) \qquad (3.76)$$

and competitive product inhibition by glucose

$$(E_\beta) + (P_1) \rightleftarrows (P_1 E_\beta). \qquad (3.77)$$

This mechanism is depicted in Fig. 3.21. The effective portion of adsorbed enzyme, ESE, at the surface of cellulose particles is assumed to be responsible for the decomposition of cellulose, Moreover, cellobiose is assumed to be converted to glucose in the aqueous phase by β-glucosidase.

The maximum or initial hydrolysis rate, R_0, is considered to depend on the initial extent of soluble protein (enzyme) adsorption, $(ES)_0$, and the effectiveness of the adsorbed soluble protein to promote the hydrolysis, k_0, as

$$R_0 = k_0 (ES)_0. \qquad (3.78)$$

The initial extent of soluble protein adsorption is related to the initial cellulose concentration, $(S)_0$, enzyme concentration, $(E)_0$, and the specific surface area of

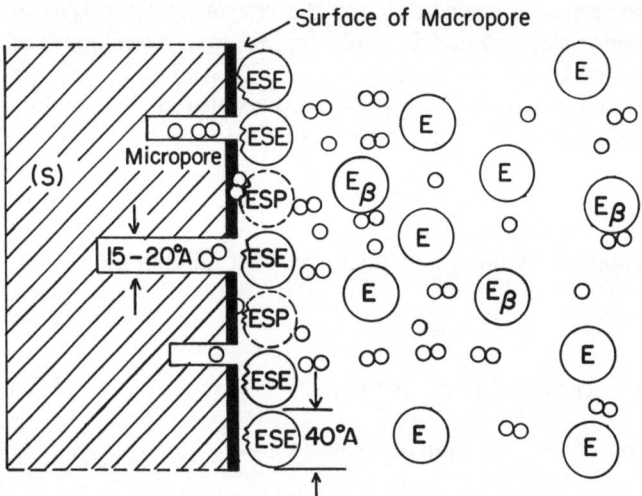

Fig. 3.21. Graphical representation of the proposed mechanism of enzyme hydrolysis of cellulose [63]

cellulose, $(SSA)_0$, through the expression

$$(ES)_0 = \frac{v_s (S)_0 (E)_0}{k_s + (S)_0} + b (SSA)_0, \qquad (3.79)$$

where v_s, k_s, and b are constants. The effectiveness of the initially adsorbed soluble protein, k_0, is related to the initial crystallinity index, $(CrI)_0$ according to the following relationship;

$$k_0 = \alpha - \beta (CrI)_0, \qquad (3.80)$$

where α and β are constants.

It is assumed that the decline in the hydrolysis rate is attributable to the change in the extent of soluble protein adsorption, the inactivation of the adsorbed enzyme, and the transformation of cellulose into a less digestible form. Suppose that the intrinsic extent of soluble protein adsorption based on unit weight of residual cellulose, (es), is almost constant throughout the hydrolysis; then,

$$(es) = (es)_0 = (ES)_0/(S)_0. \qquad (3.81)$$

The apparent extent of soluble protein adsorption based on unit volume of cellulose suspension, (ES), decreases in proportion to the concentration of residual cellulose, (S), as

$$(ES) = (es)_0 (S) = \frac{(ES)_0}{(S)_0} (S) \qquad (3.82)$$

furthermore, some portion of the adsorbed soluble protein, (ES) is converted into the ineffective form, (ESP), in the presence of the products, mainly glucose and

100

cellobiose. This mechanism is the same as the so-called non-competitive inhibition, where some portion of the adsorbed enzyme combines with the products and then is converted into the ineffective enzyme-substrate-product complex, (ESP). The effective fractions of adsorbed enzyme, θ, under the presence of different glucose and cellobiose concentrations, are expressed for cellobiose, (P_2), as

$$\theta = \frac{(ES)^e}{(ES)} = e^{-k_{P_2}(P_2)} \tag{3.83}$$

and for glucose, (P_1),

$$\theta = \frac{(ES)^e}{(ES)} = 1 - k_{P_1}(P_1), \tag{3.84}$$

where $(ES)^e$ is the effective portion of the extent of soluble protein adsorption and k_{P_1} and k_{P_2} are constants. The enzymatic hydrolysis of cellulose usually yields glucose and cellobiose simultaneously; therefore, the overall θ is expressed as a product of each θ, i.e.,

$$\theta = \frac{(ES)^e}{(ES)} = \{1 - k_{P_1}(P_1)\}\{e^{-k_{P_2}(P_2)}\} \tag{3.85}$$

and thus,

$$(ES)^e = (ES)\{1 - k_{P_1}(P_1)\}\{e^{-k_{P_2}(P_2)}\}. \tag{3.86}$$

If the structural transformation into a less digestible form does not occur, the hydrolysis rate is proportional to the effective portion of adsorbed soluble protein, that is,

$$R_0 = k_0 (ES)^e. \tag{3.87}$$

However, the hydrolysis rate further decreases because the transformation of cellulose into a less digestible form occurs, therefore, this aspect is also included in the above rate expression. The degree of structural transformation is measured in terms of the relative digestibility of cellulose, ϕ, which is defined as the ratio between the hydrolysis rate of residual cellulose and that of fresh cellulose. ϕ is related to the degree of conversion of cellulose, X, as

$$\phi = \frac{(S)^e}{(S)} = 1 - X^n = 1 - \left\{\frac{(S)_0 - (S)}{(S)_0}\right\}^n, \tag{3.88}$$

where $(S)^e$ is the effective (or active) form of cellulose and n is a parameter which is related to the structural transformation of cellulose. Combination of Eqs. (3.86)–(3.88) yields the rate expression for the decomposition of cellulose into the soluble product, mainly cellobiose, i.e.,

$$R = \frac{d(P)}{dt}$$

$$= -\frac{d(S)}{0.9\,dt} = k_0\,\phi\,(ES)^e$$

$$= k_0\left[1 - \left\{\frac{(S)_0 - (S)}{(S)_0}\right\}^n\right]\left\{\frac{(ES)_0}{(S)_0}(S)\right\}\{1 - k_{P_1}(P_1)\}\{e^{-k_{P_2}(P_2)}\}. \tag{3.89}$$

The conversion of cellobiose to glucose by β-glucosidase is competitively inhibited by glucose as depicted in Eqs. (3.76) and (3.77); thus,

$$R_{P_1} = \frac{d(P_1)}{dt} = \frac{V_m(P_2)}{K_m + (P_2) + \dfrac{K_m(P_1)}{K_I}}. \tag{3.90}$$

The reducing sugar concentration, (P), and the glucose concentration, (P_1), are obtained by integrating Eqs. (3.89) and (3.90), respectively. For simplicity, cellobiose and glucose are assumed to be the major products. The cellobiose produced, (P_2), found from the stoichiometric relationship

$$(P_2) = (P) - (P_1). \tag{3.91}$$

The residual cellulose concentration, (S), is calculated from the reducing sugar concentration, measured in terms of glucose, as

$$(S) = (S)_0 - 0.9\,(P). \tag{3.92}$$

The parameter 0.9 in the above expression is the correction factor for water taken up by the product, glucose, during hydrolysis. The product and substrate concentrations are expressed in mass units. The extent of unadsorbed soluble protein in the supernatant, (EF), is

$$(EF) = (E)_0 - (ES) = (E)_0 - (es)_0\,(S). \tag{3.93}$$

This model is useful not only for predicting the progress of hydrolysis but also for estimating changes in parameters including the extent of soluble protein adsorption, the extent of unadsorbed soluble protein, and the digestibility of cellulose. Knowledge of these parameters is essential in designing a practical hydrolysis process for enzymatic hydrolysis of insoluble cellulose.

3.4.2.8 Miscellaneous Models

To understand the mechanism of hydrolytic reaction, the influence of an individual structural feature of cellulose on the hydrolysis rate need be investigated. To facilitate it, the influence of each particular structural feature must be considered one at a time. Unfortunately, altering one structural feature of native lignocellulosics results in substantial changes in others. For example, physical pretreatments for modifying the crystallinity of lignocellulosics also alter the surface area. Similarly, chemical pretreatments substantially change all the important structural features simultaneously. Obviously the structural features are interrelated. Due to this interaction of structural features, it is desirable to establish a relationship that takes into account simultaneously the effects of the majority of the structural features on the hydrolysis rate. The following models have been proposed by Gharpuray et al. [75].

$$REH = a_1 (SSA) + a_2 (100 - CrI) + a_3 (Lignin) + a_4 \qquad (3.94)$$

and

$$\log REH = b_1 \log (SSA) + b_2 \log (100 - CrI) + b_3 \log (Lignin) + b_4, \quad (3.95)$$

where REH represents the relative extent of hydrolysis, SSA represents the specific surface area of the substrate, and Lignin represents the lignin content of the substrate. The relative extent of hydrolysis is defined as the ratio of reducing sugar concentration after 8 h of hydrolysis for pretreated substrate to that for unpretreated substrate. The model's parameters have been estimated from the experimental data. The resultant value of the parameters indicate that according to both models, (i) the specific surface area is more closely related to the relative extent of hydrolysis than the other structural features, (ii) increasing the crystallinity index diminishes the relative extent of hydrolysis, and (iii) increasing the lignin content decreases the relative extent of hydrolysis.

The relationship represented by the second model [Eq. (3.95)] is selected due to its ability to predict experimental results obtained with wheat straw [75]. The model upon rearrangement yields the following form

$$REH = 2.044 (SSA)^{0.988} (100 - CrI)^{0.257} (Lignin)^{-0.388} \qquad (3.96)$$

which is shown in Fig. 3.22. This expression implies that an increase in the surface area and a reduction in the crystallinity index and lignin content enhance the relative extent of hydrolysis. The specific surface area is the most influential structural feature, which is followed by the lignin content, which, in turn, is followed by the crystallinity index. The effects of the crystallinity index and lignin content might be embedded in the effect of specific surface area to some extent due to interaction among the structural features.

It has been reported [67] that the rate of enzymatic hydrolysis of pure cellulose tends to increase with an increase in the specific surface area and a decrease in the crystallinity index. The extent of hydrolysis after 8 h (g reducing sugar/l), X_8, is expressed as a function of both the specific surface area and crystallinity index. The following empirical expression results from a linear regression analysis of the available experimental data [67].

$$X_8 = 0.380 (SSA)^{0.195} (100 - CrI)^{1.04}. \qquad (3.97)$$

This expression indicates that the rate of hydrolysis is more sensitive to the crystallinity index. Naturally, inclusion of additional parameters, e.g., the degree of polymerization or extent of depolymerization of cellulose may improve this correlation.

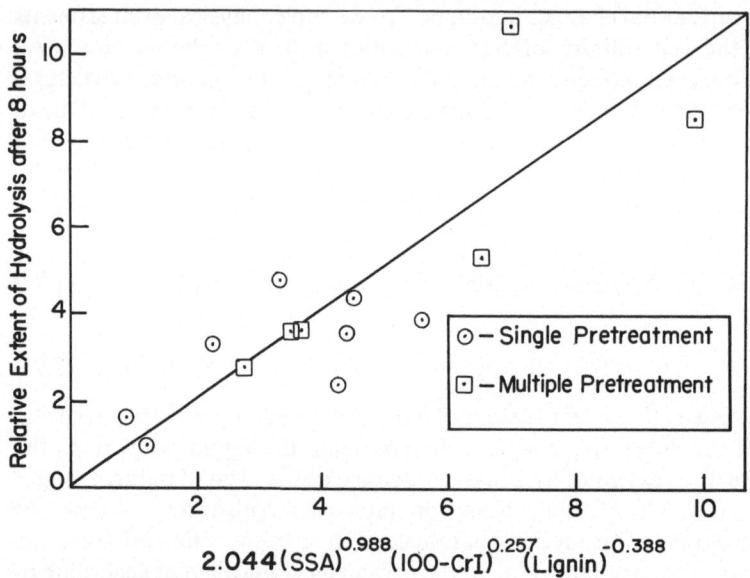

Fig. 3.22. The relative extent of hydrolysis after 8 h as a function of the specific surface area, crystallinity index, and lignin content of wheat straw [75]

3.4.3 Kinetics of Cellulases on β-1,4-Oligoglucosides

Investigations were undertaken by Whitaker [305, 306] to study the degradation of the oligoglucosides containing two, three, four, five, and six glucose units by *Myrothecium verrucaria* cellulase. The rates of hydrolysis and the sequences of products formed were compared. Table 3.16 shows the turnover numbers and rate constants at different initial concentrations of each different cellooligosaccharide obtained. Turnover numbers reveal two marked trends. First, the turnover numbers increase with an increasing degree of polymerization (DP) of the substrate up to at least a DP of five. The rate of increase is particularly high when the DP is between two and four. Second, the turnover numbers are inversely proportional to the initial substrate concentration. The initial velocity constants reflect the same trend as the turnover numbers.

A theoretical treatment of the kinetics of degradation of long chain polymers was given by Simha [260]. To test the applicability of this type of analysis, Whitaker [306, 310] compared his experimental results with the calculated values from the following rate equations:

$$\frac{dG^4}{dt} = -k^4 G^4 = -(k_{2:2}^4 + k_{3:1}^4)\, G^4, \tag{3.98}$$

$$\frac{dG^3}{dt} = \frac{3}{4}\, k_{3:1}^4\, G^4 - k^3 G^3, \tag{3.99}$$

104

Table 3.16. Initial turnover numbers and rate constants at constant enzyme concentration [306]

Substrate	Initial conc.	Turnover number	Rate constants	
			Zero-order mole ml^{-1} min^{-1}	First-order
	mM		$\times 10^9$	$min^{-1} \times 100$
Cellobiose	28.2	5.1	1.6	
	14.4	5.3	1.7	
	7.4	6.0	1.9	
	3.7	6.2	2.1	
Cellotriose	15.5	210	67	0.50
	10.2	129	41	0.48
Cellotetraose	4.0	49	16	0.45
	14.3	377	121	0.98
	7.4	316	102	1.78
	3.6	384	123	4.75
Cellopentaose	8.4	460	147	
	5.3	435	139	
	2.7	473	152	
Cellohexaose	4.5	430	138	
	2.2	439	141	

$$\frac{dG^2}{dt} = k_{2:2}^4 G^4 + \frac{2}{3} k^3 G^3 - k^2 G^2, \tag{3.100}$$

where

G^2, G^3, G^4 = concentration of cellobiose, cellotriose, cellotetraose, respectively.
k^3, k^4 = first order reaction rate constants for cellotriose and cellotetraose, respectively.

To simplify the form of the solution, the rate constant for cellobiose, k^2, is taken to be that of a zero-order reaction. Under the condition at $t = 0$, $G^4 = G_0^4$, $G^3 = G^2 = G^1 = 0$ and $k^4 \neq k^3$, solutions of Eqs. (3.98)–(3.100) are:

$$G^4 = G_0^4 e^{-k^4 t}, \tag{3.101}$$

$$G^3 = \frac{3}{4} \left[\frac{k_{3:1}^4}{k^4 - k^3} \right] G_0^4 (e^{-k^3 t} - e^{-k^4 t}), \tag{3.102}$$

$$G^2 = \frac{k_{2:2}^4}{k^4} G_0^4 (1 - e^{-k^4 t}) + \frac{k_{3:1}^4}{2(k^4 - k^3)} G_0^4 (1 - e^{-k^3 t})$$
$$- \frac{k_{3:1}^4 k^3}{2 k^4 (k^4 - k^3)} G_0^4 (1 - e^{-k^4 t}) - k^2 t, \tag{3.103}$$

$$G^1 = G_0^4 - G^4 - G^3 - G^2, \tag{3.104}$$

$$\sum G_i = G_0^4. \tag{3.105}$$

105

The calculated values agree reasonably well with the experimental values, which suggests that the conditions expressed by the rate equations were approximately satisfied during the degradation considered.

Later Whitaker [308, 310] attempted to define the kinetic behavior of the action of cellulase on β-1,4-oligoglucosides. He has found that, at a sufficiently low cellulase concentration, (E), and a high substrate concentration, (S), the rate of hydrolysis is directly proportional to the concentrations of both the cellulase and substrate, i.e.,

$$-\frac{dS}{dt} = k(E)(S), \tag{3.106}$$

where k is the second order rate constant. When (E) is constant, this expression reduces to

$$-\frac{dS}{dt} = k'(S), \tag{3.107}$$

where k′ is the first order or, strictly speaking, the pseudo-first order rate constant. The rate constants obtained are listed in Table 3.17. He has also observed that enzymatic hydrolysis of β-1,4-oligoglucosides is inhibited by increasing the concentrations of the substrate irrespective of the chain length of the latter, as shown in Fig. 3.23. The inhibition by the substrate might have resulted from an ESS complex formation. A similar observation was obtained by Nisizawa et al. [190], where cellulase components from the culture filtrate of *Irpex lacteus* have been employed.

Table 3.18 contains the Michaelis constants (K_m) of the endo-glucanase and exo-glucanase from *T. viride* for the cellodextrin series, estimated by Li et al. [155]. The Michaelis constants (K_m) of exo-glucanase are much smaller than those of the endo-glucanase for all polymer series, cellobiose through cellohexaose. This means that the exo-glucanase, which successively removes single glucose units from the non-reducing end of a cellulose chain, is more like β-glucosidase than endo-glucanase.

Table 3.17. Pseudo-first order rate constants of the hydrolysis of β-1,4-oligoglucosides by *Myrothecium* cellulase (20 µg of protein in ml) [311]

Substrate	K′, $s^{-1} \times 10^5$
Cellobiose	1.2
Cellotriose	16
Cellotetraose	83
Cellopentaose	500

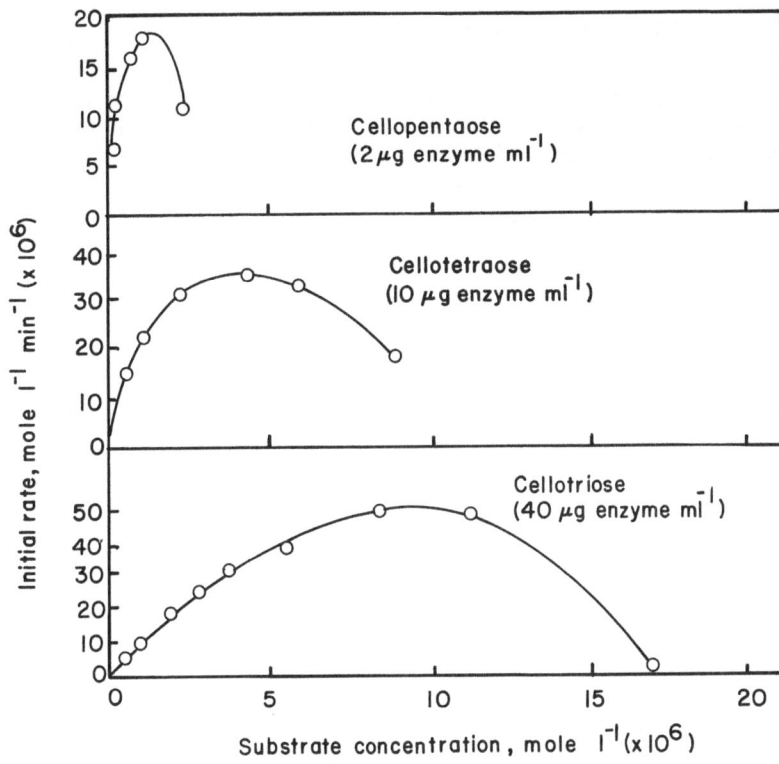

Fig. 3.23. Rates of hydrolysis of β-1,4-oligoglucosides by *Myrothecium* cellulase with increasing concentrations of substrates [310]

Table 3.18. Michaelis constants for cellulose polymer series endo- and exo-glucanase [155]

Substrate	Michaelis constant	
	Endo-glucanase	Exo-glucanase [b]
Cellobiose	$190 \times 10^{-4}\,M$ [a]	$220 \times 10^{-5}\,M$ [a]
Cellotriose	$31 \times 10^{-4}\,M$	$18 \times 10^{-5}\,M$ [a]
Cellotetraose	$28 \times 10^{-4}\,M$	$6.5 \times 10^{-5}\,M$ [a]
Cellopentaose	$7.0 \times 10^{-4}\,M$	$6.0 \times 10^{-5}\,M$
Cellohexaose	$1.0 \times 10^{-4}\,M$ [a]	$16.0 \times 10^{-5}\,M$ [a]

[a] These Lineweaver-Burk plots showed apparent substrate inhibition at concentrations exceeding 0.05 M
[b] It is not clear whether this enzyme was exo-β-1,4-glucanase (EC not given) or β-glucosidase (EC 3.2.1.21)

107

3.4.4 Kinetics of β-Glucosidase

The purification, physical characteristics and mechanism of cellobiase from *T. viride* were examined by Gong et al. [84]. Based on analysis of the initial rate data, plotted in the form of the Lineweaver-Burk, they concluded that cellobiase is non-competitively inhibited by the product glucose. Therefore, the reaction sequence for this enzyme could be written as

$$E + GG \underset{k_2}{\overset{k_1}{\rightleftharpoons}} E^* \overset{k_3}{\longrightarrow} E + 2G, \tag{3.108}$$

$$E^* + G \underset{k_6}{\overset{k_5}{\rightleftharpoons}} E^*G, \tag{3.109}$$

$$E + G \underset{k_8}{\overset{k_7}{\rightleftharpoons}} EG, \tag{3.110}$$

where

E = free enzyme
GG = cellobiose
G = glucose.

By applying the pseudo-steady state assumption, Gong et al. [84] derived the inverted form of the kinetic expression.

$$\frac{1}{V} = \frac{K\,[1+(G)/K_{i,2}]}{V} \cdot \frac{1}{(GG)} + \frac{1+(G)/K_{i,1}}{V}, \tag{3.111}$$

where

$$K = \frac{k_2 + k_3}{k_1}$$

$$K_{i,1} = \frac{k_6}{k_5}$$

$$K_{i,2} = \frac{k_8}{k_7}$$

$$V = 2k_3\,E_{tot}.$$

Since the plots and replots are linear, values of the kinetic constants, K, $K_{i,1}$, and $K_{i,2}$ can be determined directly from them. The kinetic constants for three cellobiase fractions are summarized in Table 3.19. The kinetic equation derived from the proposed mechanism accurately predicted the time course of hydrolysis of cellobiose by cellobiase over an eightfold range of the substrate concentration and at conversions of over 90 % [84]. The results indicate that cellobiase from *T. viride* consists of several chromatographically distinct and yet kinetically similar fractions, which are subject to product inhibition, and that these fractions of cellobiase inhibit hydrolysis of cellobiose according to a non-competitive mechanism.

Table 3.19. Values of kinetic constants for peaks 1, 2, and 3 cellobiase fractions [84]

Kinetic constant	Cellobiase enzyme fraction		
	Peak 1	Peak 2	Peak 3
$\dfrac{V\ \mu\text{mol glucose}}{E_{\text{tot}}\ \text{min}\cdot\text{mg}^{-1}\ \text{protein}}$	66.2	116.0	44.6
K (mM)	2.65	2.5	2.74
$K_{i,1}$ (mM)	–	16.4	14.7
$K_{i,2}$ (mM)	–	1.22	4.26

Nomenclature

a	= radius of enzyme, cm
A_s	= surface area of a substrate, L^2 [Eqs. (3.29) and (3.31)]
b	= constant [Eq. (3.68)]
b	= constant, dimensionless
C	= concentration of product, ML^{-3} [Eqs. (3.57) and (3.60)]
C_A	= concentration of amorphous cellulose, ML^{-3} [Eq. (3.46)]
C_C	= concentration of crystalline cellulose, ML^{-3} [Eq. (3.47)]
C_E	= concentration of enzyme in the particle, mol cm^{-3} [Eq. (3.63)]
C_E	= concentration of free enzyme [Eqs. (3.2) and (3.62)]
C_{ES}	= concentration of enzyme-substrate complex, ML^{-3} [Eq. (3.2)]
C_{Et}	= total enzyme concentration, ML^{-3} [Eqs. (3.4) and (3.53)]
C_i	= cellulose concentration of component i ML^{-3} [Eq. (3.15)]
C_i	= substrate fragments of chain length i, mol cm^{-3} [Eqs. (3.2) and (3.62)]
C_i	= concentration of cellulosic polymer with degree of polymerization equal to i [Eq. (3.53)]
CrI	= crystallinity index (dimensionless) [Eq. (3.80)]
C_s	= concentration of substrate in the particle, mol cm^{-3}
C_{sL}	= concentration of substrate in the bulk solution, mol cm^{-3} [Eq. (3.69)]
$C_I,\ C_{II}$	= concentrations in each phase, ML^{-3} [Eq. (3.30)]
D	= diffusion constant, cm^2 s^{-1} [Eq. (3.56)]
D	= inactivated form of the enzyme-substrate complex, ML^{-3} [Eq. (3.27)]
D_s	= average restricted diffusion coefficient of the soluble substrate fragments, cm^2 s^{-1} [Eq. (3.62)]
D_{SF}	= average free diffusion coefficient of the fragments, m^2 s^{-1} [Eq. (3.67)]
e	= enzyme concentration, ML^{-3} [Eq. (3.12)]
E	= enzyme [Eqs. (3.1), (3.24), (3.34), (3.35), (3.52) and (3.61)]
E	= enzyme concentration in supernatant (g/L) [Eq. (3.29)]
E^*	= enzyme-substrate complex concentration, ML^{-3}
E_a	= enzyme adsorbed on the surface, mol [Eq. (3.31)]
E_A^A	= enzyme fraction active on amorphous cellulose, ML^{-3} [Eqs. (3.46) and (3.50)]
E_A^C	= enzyme fraction active on crystalline cellulose, ML^{-3} [Eqs. (3.47) and (3.51)]
E_A^d	= deactivated form of enzyme
E_{ads}	= adsorbed protein, (mg protein) (mg cellulose^{-1}) [Eq. (3.10)]
$E_{ads,m}$	= maximum adsorbed protein, (mg protein) (mg cellulose^{-1}) [Eq. (3.10)]
EF	= apparent extent of unadsorbed soluble protein (g/L)
E_I	= number of moles of enzyme on the surface, mol [Eq. (3.31)]
E_0	= protein concentration in the supernatant, mg ml^{-1} [Eq. (3.10)]
(E_0)	= initial enzyme concentration, ML^{-3} [Eqs. (3.28) and (3.37)]
EP	= enzyme-product complex (G/L)

109

ES	= apparent extent of soluble protein adsorption (g/L)
es	= intrinsic extent of soluble protein adsorption (g/g) [Eq. (3.81)]
$ES^e = ESE$	= effective portion of (ES) (g/L) [Eqs. (3.73), (3.75), (3.85), and (3.86)]
ESP	= ineffective portion of (ES) or cellulase-cellulose-reducing-sugar complex (g/L)
ESP_1	= ineffective portion of (ES) or cellulase-cellulose-glucose complex (g/L) [Eq. (3.72)]
ESP_2	= ineffective portion of (ES) or cellulase-cellulose-cellobiose complex (g/L) [Eq. (3.71)]
E_β	= β-glucosidase [Eq. (3.77)]
$E_\beta P_2$	= β-glucosidase-cellobiose complex
E_1	= enzyme concentration (exo-glucanase), ml^{-3} [Eq. (3.19)]
$E*G$	= enzyme-substrate-product complex, ML^{-3}
EP	= enzyme-product complex [Eq. (3.26)]
ES	= enzyme-substrate complex [Eqs. (3.1), (3.24), (3.27), and (3.61)]
f	= fraction of amorphous cellulose in the total cellulose, dimensionless [Eq. (3.43)]
G	= glucose concentration, ML^{-3} [Eq. (3.108)]
G_2	= cellobiose concentration [Eqs. (3.19) and (3.23)]
$[G_2]$	= cellobiose in polymerized form [Eqs. (3.19) and (3.23)]
$[G_2]_0$	= initial cellobiose concentration in polymerized form, mol^{-3} [Eq. (3.23)]
G_1, G_2, G_3, G_4	= glucose, cellobiose, cellotriose, cellotetraose concentration, ML^{-3} [Eqs. (3.98), (3.99), and (3.100)]
G_x	= reducing sugar concentration, ML^{-3} [Eqs. (3.48) and (3.49)]
GG	= cellobiose concentration, ML^{-3} [Eq. (3.108)]
[I]	= inhibitor concentration, ML^{-3}
k_{p_1}, k_{p_2}	= constants [Eqs. (3.83) and (3.84)]
k_s, v_s	= constants
k_s, v_s	= constants
k	= rate constant, T^{-1} [Eqs. (3.12), (3.13), and (3.14)]
k', k''	= rate constants, T^{-1}
K	= Michaelis constant [Eq. (3.11)]
K	= partition coefficient [Eqs. (3.30) and (3.33)]
k_1	= reaction rate constant for formation of enzyme-substrate complex [Eqs. (3.1), (3.34), (3.52), and (3.61)]
k_2	= reaction rate constant for decomposition of enzyme-substrate complex into enzyme and substrate [Eqs. (3.1), (3.25), (3.52), and (3.61)]
k_3	= reaction rate constant for decomposition of enzyme-substrate complex into enzyme and product [Eqs. (3.1), (3.35), (3.52), and (3.61)]
k_i	= rate constant associated with cellulose component i, T^{-1} [Eq. (3.15)]
K_i	= dissociation constant for the EP complex, ML^{-3} [Eq. (3.22)]
K_i'	= modified equilibrium constant between enzyme and products [Eq. (3.18)]
K_m	= Michaelis constant, ML^{-3} [Eqs. (3.6) and (3.53)]
K_m'	= modified Michaelis constant, gl^{-1} [Eqs. (3.18) and (3.55)]
K_p	= constant, $ml\,mg^{-1}$ [Eqs. (3.10) and (3.11)]
K_p	= equilibrium constant between enzyme and products
K_s	= dissociation constant for the ES complex [Eq. (3.22)]
$\bar{k}_s$	= overall mass transfer coefficient of substrate, $cm\,s^{-1}$ [Eq. (3.69)]
$k_1, k_{-1}, k_2,$ k_3, k_4, k_5, k_{-5}	= rate constants, T^{-1} [Eqs. (3.27), (3.28), (3.98), (3.99), and (3.100)]
K_1	= constant, dimensionless
K_3	= constant, dimensionless
K_5	= constant, dimensionless
k^2, k^3, k^4	= rate constants for cellobiose, cellotriose and cellotetraose
$k_{2:2}, k_{3:1}$	= rate constants for cellobiose, cellotriose and cellotetraose [Eq. (3.98)]
m	= constant, dimensionless [Eq. (3.12)]
M	= mass taken up at the boundary, ML^{-2} [Eq. (3.60)]
n	= parameter for the structural transformation of cellulose (dimensionless)
n	= constant, dimensionless [Eqs. (3.12) and (3.14)]
n	= number of particles [Eq. (3.69)]

O	= subscript referring to initial values
P	= product, (g/l) [Eqs. (3.1), (3.35), (3.36), and (3.61)]
P	= product concentration, ML^{-3} [Eqs. (3.11), (3.15), (3.25), (3.26), and (3.28)]
$\hat{P}$	= $(P) - (P)_0$ (g/L)
P_1	= glucose concentration (g/l) [Eq. (3.72)]
$\hat{P}_1$	= $(P_1) - (P_1)_0$ (g/L)
P_2	= cellobiose concentration (g/L) [Eq. (3.71)]
$\hat{P}_2$	= $(P_2) - (P_2)_0$ (g/L)
P, (P)	= product concentration, ML^{-3} [Eqs. (3.22) and (3.69)]
r	= intrinsic hydrolysis rate (h^{-1})
r	= radial position within particles, cm
R	= apparent hydrolysis rate (g/L h) [Eq. (3.78)]
R	= particle radius, cm [Eq. (3.69)]
R_{p_1}	= apparent glucose production rate (g/L h)
S	= substrate [Eqs. (3.1), (3.34), (3.52), and (3.61)]
S	= cellulose concentration (g/L) [Eq. (3.24)]
S^e	= cellulose concentration in digestible form (g/L)
S^i	= cellulose concentration in undigestible form (g/L) [Eq. (3.74)]
SSA	= specific surface area (m^2/g) [Eq. (3.79)]
(S), [S]	= cellulose concentration, ML^{-3} [Eqs. (3.11) and (3.37)]
$[S]_a$	= effective substrate concentration, gl^{-1} [Eq. (3.16)]
S_a	= amorphous cellulose concentration, ML^{-3} [Eq. (3.40)]
S_c	= crystalline cellulose concentration, ML^{-3} [Eq. (3.40)]
$[S]_i$	= substrate concentration at time i, gl^{-1} [Eqs. (3.17) and (3.18)]
$[S]_{i+t_r}$	= substrate concentration at time $i + t_r$, gl^{-1} [Eq. (3.17)]
$[S]_t$	= total substrate concentration, gl^{-1} [Eq. (3.16)]
S_0, $(S)_0$, $(S)_0$, $[S]_0$	= initial cellulose concentration, gl^{-1} or $mol\,l^{-1}$ [Eqs. (3.18), (3.22), (3.28), and (3.38)]
t	= hydrolysis time (h) [Eqs. (3.12), (3.13), (3.14), (3.22), (3.23), and (3.39)]
t	= time, T
t_r	= time required for reducing cellulose concentration from $[S]_i$ to $[S]_{i+t_r}$ [Eqs. (3.17) and (3.18)]
T	= rate of reaction, $ML^{-3}T^{-1}$
V	= maximum rate of reaction, $ML^{-3}T^{-1}$ [Eqs. (3.11) and (3.22)]
v	= reaction rate, $ML^{-3}T^{-1}$ [Eqs. (3.11), (3.37), and (3.39)]
V′	= modified maximum rate of reaction, $gl^{-1}h^{-1}$ [Eqs. (3.17) and (3.18)]
v_i	= initial reaction rate, $ML^{-3}T^{-1}$
V_I	= volume of phase I at equilibrium, L^3 [Eq. (3.31)]
V_L	= volume of the liquid phase, cm^3 [Eq. (3.69)]
v_m, K_m, K_1	= constants
WRV	= water retention volume (g/g)
X	= degree of conversion of cellulose (dimensionless) [Eq. (3.88)]
x	= distance normal to surface of reaction, L [Eq. (3.56)]
X	= extent of hydrolysis, % [Eqs. (3.12), (3.13), and (3.14)]
X_A, X_c	= enzyme-substrate complex [Eqs. (3.48) and (3.49)]
X_1, X_2, X_3	= enzyme-substrate-product complex
X_{lm}	= maximum value of X_1 [Eqs. (3.37) and (3.39)]
X_2	= enzyme-substrate complex [Eq. (3.36)]
X_1	= enzyme-substrate complex [Eq. (3.34)]
X_1	= enzyme-crystalline cellulose complex [Eq. (3.40)]
X_2	= enzyme-amorphous cellulose complex [Eq. (3.41)]
X_3	= enzyme-product complex [Eq. (3.42)]
Y_A, Y_c	= enzyme-substrate-product complexes

Greek Letters

α, β = constants [Eq. (3.80)]
α = constant, dimensionless [Eq. (3.16)]
μ_n = number average molecular weight [Eq. (3.9)]
θ = fraction of $(ES)^e$ (dimensionless) [Eq. (3.83)]
ϕ = relative digestibility of residual cellulose (dimensionless) [Eq. (3.89)]
γ = pore radius of particle, cm [Eq. (3.66)]
η = correction factor pertaining to diffusional restriction, dimensionless
ϕ = constant, dimensionless [Eq. (3.66)]

References

1. Anonymous (1979) Continuous cellulose to glucose process. Chem Eng News 57 (41):19
2. Anonymous (1980) A fungus to digest Lignin. Gasohol USA 2 (10):26
3. Aggebrandt LG, Samuelson OJ (1964) J Appl Poly Sci 8:2801
4. Ahlgren E, Ericksson KE (1969) Acta Chem Scand 21:1193
5. Almin KE et al. (1975) Eur J Biochem 51:207
6. Amemura A, Terui G (1965a) J Ferment Technol 43:275
7. Amemura A, Terui G (1965b) ibid. 43:281
8. Ander P, Ericksson KE (1978) In: Bull MJ (ed) Progress in ind microbiol, vol 14. Elsevier, New York, p 1
9. Anderson AW, Han YW (1977) NSF-RA-770278, Non Conventional Protein Food Processing Conference, PB-283965, p 202
10. Anderson DC, Ralston AT (1973) J Anim Sci 37:148
11. Andren RK, Nystrom JM (1976) AIChE Symp Ser 72 (158):91
12. Andren RK et al. (1975) In: Timell TE (ed) Appl Poly Symp no 28. Interscience, New York, p 205
13. Andren RK et al. (1976) In: Gaden EL Jr et al. (eds) Biotechnol Bioeng Symp no 6. Interscience, New York, p 177
14. April GC et al. (1980) Presented at the 89th AIChE National Meeting, Portland, Oregon, August 17–20
15. Arminger WG et al. (1976) AIChE Symp Ser 72 (158):77
16. Arthur JC Jr (1968) In: Phillips GO (ed) Energetics and mechanisms in radiation biology. Academic Press, New York
17. Autrey KM et al. (1975) J Dairy Sci 58:67
18. Bailey M et al. (eds) (1975) Proc Symp Enz Hydrol Cellulose. SITRA, Helsinki
19. Baker AJ (1973) J Anim Sci 36:768
20. Baker TL et al. (1959) ibid. 18:655
21. Barras DR et al. (1969) Adv Chem Ser 95:105
22. Bastawde KB et al. (1978) In: Ghose TK (ed) Bioconversion of cellulosic substances into energy chemicals and microbial protein Symp. BERC IIT Delhi, New Delhi, p 387
23. Beardmore DH et al. (1980) Biotechnol Lett 2:435
24. Bender F et al. (1970) Forest Prod J 20:36
25. Berghem LER, Pettersson LG (1973) Eur J Biochem 27:21
26. Berghem LER et al. (1975) ibid. 53:55
27. Berghem LER et al. (1976) ibid. 61:621
28. Blouin FA, Arthur JC Jr (1960) J Chem Eng Data 5:470
29. Borrevik RK et al. (1978) In: Effect of nitrogen oxide pretreatment on enzymatic hydrolysis of cellulose, LBL 7879, September 1978. Lawrence Berkeley Lab., Univ. of Calif, Berkeley
30. Bowers GH, April GC (1977) TAPPI 60 (8):102
31. Brandt D et al. (1973) AIChE Symp Ser 69 (133):127
32. Brenner W et al. (1977) In: New approaches for the acid hydolysis of cellulose. Dept. of Appl. Sci., New York Univ., NYU/DAS-77-30, New York
33. Brink DL (1978) US Patent 4, 076, 579, Feb.
34. Brink DL et al. (1974) In: Abstracts of papers of the 167th Amer. Chem. Soc. Meeting, Los Angeles, Ca. March 31–April 5

35. Brown DE, Waliuzzaman M (1977) In: Ghose TK (ed) Proc Bioconversion Symp. BERC ITT Delhi, New Delhi, p 351
36. Browning BL (ed) (1963) The chemistry of wood. Interscience, New York, p 429
37. Casey JP (1960) Pulp and paper, vol 1. Interscience, New York
38. Caulfield DF, Moore WE (1974) Wood Science 6:375
39. Chandra S, Jackson MG (1971) J Agr Sci Camb 77:11
40. Clarke SD, Dyer IA (1973) J Anim Sci 37:1022
41. Cowling EB (1963) In: Reese ET (ed) Advances in enzymatic hydrolysis of cellulose and related material. Pergamon, New York, p 1
42. Cowling EB (1975) In: Wilke CR (ed) Biotechnol Bioeng Symp, no 5. Interscience, New York, p 163
43. Cowling EB, Brown W (1969) Adv Chem Ser 95:152
44. Cowling EB, Kirk TK (1976) In: Gaden EL Jr et al. (eds) Biotechnol Bioeng Symp, no 6. Interscience, New York, p 95
45. Datta R (1980) Presented at AIChE 73rd Annual Meeting, Chicago, Nov. 16–20
46. Dehority BA, Johnson JJ (1961) J Dairy Sci 44:2242
47. Dellweg H (ed) (1976) Proc V Intern Ferment Symp, Berlin
48. Detroy RW et al. (1980) In: Scott CD (ed) Biotechnol Bioeng Symp, no 10. Interscience, New York, p 135
49. Donefer E et al. (1969) Adv Chem Ser 95:328
50. Dunlap CE, Chiang LC (1980) In: Shuler MC (ed) Cellulose degradation – a common link. CRC Press, New York
51. Dunlap CE et al. (1976) AIChE Symp Ser 72 (158):58
52. Dunning JW, Lathrop EC (1945) Ind Eng Chem 37:24
53. Dwivedi CP, Ghose TK (1979) J Ferment Technol 57:15
54. Elmund GK et al. (1975) In: Management of livestock wastes, Proc 3rd Intern Symp on Livestock wastes, A.S.A.E. Publication, p 275
55. Emert GH et al. (1974) Adv Chem Ser 136:79
56. Eriksson KE (1978) In: Ghose TK (ed) Bioconversion of cellulosic substrates into Energy Chemicals and Microbial Protein Symp. Proc. BERCIIT Delhi, New Delhi, p 195
57. Eriksson KE (1975) In: Bailey M et al. (eds) Proc Symp Enz Hydrol Cellulose. SITRA, Helsinki, p 263
58. Eriksson KE, Goodell EW (1974) Can J Microbiol 20:371
59. Eriksson KE, Hollmark BH (1969) Arch Biochem Biophys 133:233
60. Eriksson KE, Pettersson B (1975a) Europ J Biochem 51:193
61. Eriksson KE, Pettersson B (1975b) ibid. 51:213
62. Eriksson KE, Rzedowski W (1969) Arch Biochem Biophys 129:689
63. Fan LT, Lee Yong-Hyun (1983) Biotechnol Bioeng 25:2707
64. Fan LT et al. (1980) In: Fiechter A (ed) Adv Biochem Eng, vol 14. Springer, Berlin Heidelberg New York, p 101
65. Fan LT et al. (1982) In: Fiechter A (ed) Adv Biochem Eng, vol 23. Springer, Berlin Heidelberg New York, p 157
66. Fan LT et al. (1980) Biotechnol Bioeng 22:177
67. Fan LT et al. (1981) ibid. 23:419
68. Fan LT et al. (1980) Presented at IInd Intern Symp Bioconv and Biochem Eng, IIT, Delhi, India, March 3–6
69. Fan LT et al. (1981) In: Scott CD (ed) Biotechnol Bioeng Symp, no 11. Wiley, New York
70. Feist WC et al. (1970) J Anim Sci 30:832
71. Ferguson WS (1942) Biochem J 36:786
72. Ferguson WS (1943) J Agr Sci 33:174
73. Gaden EL Jr et al. (eds) (1976) Biotechnol Bioeng Symp, no 6. Interscience, New York
74. Gascoigne JA, Gascoigne MM (1980) In: Biological degradation of cellulose. Butterworths, London, p 204
75. Gharpuray MM et al. (1983) Biotechnol Bioeng 25:157
76. Ghose TK (1969) ibid. 11:239
77. Ghose TK, Bisaria VS (1979) ibid. 21:131
78. Ghose TK, Das K (1971) In: Ghose TK, Fiechter A (eds) Adv in Biochem Eng, vol 1. Springer, Berlin Heidelberg New York, p 55

79. Ghose TK, Kostick JA (1969) Adv Chem Ser 95:415
80. Ghose TK et al. (1976) In: Dellweg H (ed) Proc V Intern Ferment Technol Symp. Berlin, p 439
81. Gilligan W, Reese ET (1954) Canad J Microbiol 1:90
82. Glegg RE, Kertesz ZI (1957) J Poly Sci 26:289
83. Godden W (1920) J Agr Sci 10:437
84. Gong CS et al. (1977) Biotechnol Bioeng 19:959
85. Guggolz JR et al. (1971) J Anim Sci 33:151
86. Guggolz JR et al. (1971) ibid. 33:167
87. Gum EK Jr, Brown RD Jr (1976) Biochem Biophys Acta 446:371
88. Hajny GJ, Reese ET (eds) (1969) Adv Chem Ser 95. ACS Washington, DC
89. Hall JA et al. (1956) Unasylva 10:7
90. Halliwell G (1978) In: Ghose TK (ed) Bioconversion of cellulosic substrates into energy chemicals and microbial protein. BERC IIT Delhi, New Delhi, p 81
91. Halliwell G (1961) Biochem J 79:185
92. Haliiwell G (1965) ibid. 95:270
93. Halliwell G, Griffin M (1973) ibid. 135:587
94. Halliwell G, Griffin M (1974) ibid. 2:497
95. Halliwell G, Riaz M (1970) Arch Mikrobia 116:35
96. Halliwell G, Riaz M (1971) ibid. 78:295
97. Han YW, Callihan CD (1974) Appl Microbiol 27:159
98. Han YW, Chen WP (1978) Biotechnol Bioeng 20:567
99. Han YW, Ciegler A (1982) Process Biochemistry Jan/Feb. 32
100. Han YW et al. (1975) Appl Microbiol 29:708
101. Han YW et al (1981) Biotechnol Bioeng 23:2525
102. Han YW et al. (1975) J Agr Food Chem 23:928
103. Han YW et al. (1983) ibid. Jan–Feb. 34
104. Hansen SM, April GC (1981) Biosources Digest 3 (1):Jan
105. Harris EE et al. (1969) US Dept Agr Forest Serv Rep 1618, Washington DC
106. Hash JH, King KW (1958) J Biol Chem 232:381
107. Heaney DC, Bender F (1970) Forest Prod J 20:98
108. Howell JA, Mangat M (1978) Biotechnol Bioeng 20:847
109. Howell JA, Stuck JD (1975) ibid. 17:873
110. Howsmon JA, Marchessault RH (1959) J Appl Poly Sci 1:313
111. Hsu TA et al. (1980) Chemtech, May
112. Huang AA (1975a) Biotechnol Bioeng 17:1421
113. Huang AA (1975b) In: Wilke CR (ed) Biotechnol Bioeng Symp, no 5. Interscience, New York, p 245
114. Huang AA (1975c) Unpublished Report of US. Army Natick Laboratory,
115. Huffman JG et al. (1971) Can J Anim Sci 51:457
116. Israilides CJ et al. (1979) Develop Ind Microbiol 20:603
117. Iwasaki Y et al. (1964) J Biochem 55:209
118. Jones RP et al. (1980) In: Cellulose hydrolysis, state of the art and near term potential, a report submitted to the Queensland Dept. of Commercial and Industrial Development, Jan.
119. Kanda T et al. (1976) J Biochem 79:977
120. Kanda T et al. (1976b) ibid. 79:997
121. Karrer P et al. (1926) Chim Acta 9:893
122. Karrer P et al. (1925) Helv Chim Acta 8:797
123. Kelly JA (1973) Radiolytic Hydrolysis of Cellulose, US Environ. Protec. Agency Report EPA-670/2-73-030
124. Kelsey RG, Shafizadeh F (1980) Biotechnol Bioeng 22:1025
125. Kim C (1974) In: ARO Report 74-2, Proc of the 1974 Army Numerical Analysis Conference. The Office of the Chief of Research, Development and Acquisition, p 507
126. King KW (1963) In: Reese ET (ed) Advances in enzymic hydrolysis of cellulose and related materials. Pergamon, New York, p 159
127. King KW (1967) Arch Biochem Biophys 120:462
128. King KW (1966) Biochem Biophys Res Commun 24:295

129. King KW (1965) J Ferment Technol 43:79
130. King KW (1961) Va Agric Expt Sta Tech Bull 154
131. King KW, Vessal MI (1969) Adv Chem Ser 95:7
132. Kirk TK (1975) In: Wilke CR (ed) Biotechnol Bioeng Symp, no 5. Interscience, New York, p 139
133. Kirk TK, Harkin JM (1973) AIChE Symp Ser 69 (133):124
134. Klee AJ, Rogers CJ (1977) In: Proc Second Pacific Chemical Engineering Congress, vol 2. AIChE, New York, p 759
135. Kleinert TN (1974) TAPPI 57:99
136. Knappert D et al. (1980) Biotechnol Bioeng 22:1449
137. Kobayashi T (1971) Proc Biochem 6 (2):19
138. Koenigs JW (1975) In: Wilke CR (ed) Biotechnol Bioeng Symp, no 5. Interscience, New York, p 151
139. Kooiman P et al. (1953) Enzymologia 16:237
140. Krassig J, Kitchen W (1961) J Poly Sci 51:123
141. Kumakura M, Kaetsu I (1979) Biotechnol Bioeng 20:1309
142. Kunz ND et al. (1972) Nuclear Technol 16:556
143. Ladisch MR (1977) Ph. D. Dissertation, Enzymatic Hydrolysis of Cellulose: Kinetics and Mechanism of Selected Purified Cellulose Components, Purdue University
144. Ladisch MR et al. (1978) Energy 4:263
145. Laurent TC, Killander J (1964) J Chromatogr 14:317
146. Lawton EJ et al (1951) Science 113:380
147. Lee GF Jr (1966) M. Sc. Thesis, Syracuse Univ., Syracuse N.Y.
148. Lee SE (1977) Ph. D. Dissertation, Modeling and Kinetic Studies of Cellulose Biodegradation Reactions, University of Pennsylvania, Philadelphia
149. Lee SE et al. (1978) Biotechnol Bioeng 20:141
150. Lee YH, Fan LT (1980) Presented at the VIth Intern Ferment Symp, London, Ontario, Canada, July 20–25
151. Lee YY et al. (1978) In: Scott CD (ed) Biotechnol Bioeng Symp, no 8. Interscience, New York, p 75
152. Lee, Yong-Hyun, Fan LT (1980) In: Fiechter A (ed) Adv Biochem Eng, vol 18. Springer, Berlin Heidelberg New York, p 101
153. Lee, Yong-Hyun et al. (1980) In: Fiechter A (ed) Adv Biochem Eng, vol 17. Springer, Berlin Heidelberg New York, p 131
154. Lehmann F (1905) Ger. Patent No. 169, 800, March 26
155. Li LH et al. (1985) Arch Biochim Biophys 111:439
156. Linden JC et al. (1980) Presented at the VIth Intern Ferment Symp, London, Ontario, Canada, July 20–25
157. Linko M (1977) In: Ghose TK et al. (eds) Adv Biochem Eng, vol 5. Springer, Berlin Heidelberg New York, p 27
158. Lipinsky E (1979) Adv Chem Ser 181:1
159. Liu TH, King KW (1967) Arch Biochem Biophys 120:462
160. Lloyd RA, Harris JR (1963) US Dept Agr Forest Serv Report, 2029, Washington DC
161. MacDonald DG, Mathews JF (1979) Biotechnol Bioeng 21:1091
162. Mandels M, Reese ET (1963) In: Reese ET (ed) Advances in enzymatic hydrolysis of cellulose and related material. Pergamon, New York, p 115
163. Mandels M, Reese ET (1964) Develop Indust Microbiol 5:5
164. Mandels M, Sternberg D (1976) J Ferment Technol 54:267
165. Mandels M et al. (1974) Biotechnol Bioeng 16:1471
166. Mandels M et al. (1971) J Poly Sci, part C, no 36, 445
167. Mangat MS (1977) Ph. D. Dissertation, A Kinetic Study of the Enzymatic Hydrolysis of Cellulose. State University of New York at Buffalo
168. Marsh CA (1966) Biochim Biophys Acta 122:367
169. Marsh PB (1957) J Text Res 27:913
170. Marsh PB et al. (1953) ibid. 23:831
171. McLaren AD (1963) Enzymologia 26:237
172. McLaren AD, Packer L (1970) In: Nord FF (ed) Advances in enzymology, vol 33. Interscience, New York, p 245
173. Mellenberger RW et al. (1970) J Anim Sci 30:1005

174. Mellenberger RW et al. (1971) ibid. 32:756
175. Merchant MV (1957) TAPPI 40:771
176. Millett MA (1975) In: Wilke CR (ed) Biotechnol Bioeng Symp, no 5. Interscience, New York, p 319
177. Millett MA, Goedken VL (1965) TAPPI 48:367
178. Millett MA et al. (1979) Adv Chem Ser 181:71
179. Millett MA et al. (1975) In: Wilke CR (ed) Biotechnol Bioeng Symp, no 5. Interscience, New York, p 193
180. Millett MA et al. (1976) In: Gaden EL Jr et al. (eds) Biotechnol Bioeng Symp, no 6. Interscience, New York, p 125
181. Millett MA et al. (1970) J Anim Sci 31:781
182. Miyamato S, Nisizawa K (1942) J Veterinary Soc Army Japan 396:778
183. Moore WE et al. (1972) J Agr Food Chem 20:1173
184. Moo-Young M et al. (1978) Biotechnol Bioeng 20:107
185. Nelson R, Oliver DW (1971) J Poly Sci, part C, no 36. p 305
186. Nesse N et al. (1977) Biotechnol Bioeng 19:323
187. Nisizawa K (1973) J Ferment Technol 51:267
188. Nisizawa K, Kobayashi T (1953a) J Agr Chem Soc Japan 27:239
189. Nisizawa K, Kobayashi T (1953b) ibid. 27:241
190. Nisizawa K et al. (1963) In: Reese ET (ed) Advances in enzymic hydrolysis of cellulose and related material. Pergamon, New York, p 171
191. Nisizawa K et al. (1962) Arch Biochem Biophys 96:152
192. Nisizawa K et al. (1966) J Ferment Technol 44:659
193. Nisizawa K et al. (1972) In: Terui G (ed) Proc IV Intern Ferment Symp. Osaka, p 719
194. Niwa T (1965) J Ferment Technol 43:286
195. Niwa T et al. (1964) ibid. 42:124
196. Nguyen XN et al. (1980) Presented at the 89th AIChE National Meeting. Portland, Oregon, August 17–20
197. Nolan EJ et al. (1980) Presented at the VIth Intern Ferment Symp, London, Ontario, Canada, July 20–25
198. Norkrans B (1963) Ann Rev Phytopathology 1:325
199. Norkrans B (1950) Physical Plant 3:75
200. Nystrom J (1975) In: Wilke CR (ed) Biotechnol Bioeng Symp, no 5. Interscience, New York, p 221
201. Ogawa K, Toyama N (1964) J Ferment Technol 42:199
202. Ogawa K, Toyama N (1966) ibid. 44:741
203. Ogawa K, Toyama N (1967) ibid. 45:671
204. Ogawa K, Toyama N (1968) ibid. 46:367
205. Ogawa K, Toyama N (1972) ibid. 50:236
206. Ogiwara Y, Arai K (1967) J Japan Tech Assoc Pulp Paper Ind 21:209
207. Okada G (1975) J Biochem 77:33
208. Okada G (1976) ibid. 80:913
209. Okada G, Nisizawa K (1975) ibid. 78:297
210. Okada G et al. (1968) ibid. 63:591
211. Okada G et al. (1968) J Ferment Technol 44:682
212. Okazaki M, Moo-Young M (1978) Biotechnol Bioeng 20:637
213. Ooshima H et al. (1984) Biotechnol Letters 6:289
214. Pannir Selvam PV, Ghose TK (1980) Presented at the IInd Intern Symp Bioconv Biochem Eng, IIT Delhi, India, March 3–6
215. Peitersen N (1975) In: Bailey M et al. (eds) Symp Enzymatic Hydrolysis of Cellulose. SITRA, Helsinki, p 407
216. Peitersen N, Anderson B (1978) AIChE Symp Ser 74 (172):100
217. Peitersen N et al. (1977) Biotechnol Bioeng 19:1091
218. Pettersson G (1968) Arch Biochim Biophys 123:307
219. Pettersson G, Eaker DL (1968) ibid. 124:154
220. Pettersson G, Porath J (1963) Biochim Biophys 67:9
221. Pettersson G, Porath J (1966) Methods Enzymol 8:603
222. Pettersson G et al. (1963) Biochim Biophys 67:1

223. Pettersson LG (1975) In: Bailey M et al. (eds) Proc Symp Enz Hydrol Cellulose. SITRA, Helsinki, p 255
224. Pettersson LG et al. (1972) In: Terui G (ed) Proc IV Int Ferment Symp. Osaka, p 723
225. Pew JC (1957) TAPPI 40:553
226. Pew JC, Weyna P (1962) TAPPI 45:247
227. Phillip B et al. (1979) Adv Chem Ser 181:127
228. Phillip B et al (1977) Chemtech, Nov, p 702
229. Pritchard GI et al. (1962) Can J Anim Sci 42:215
230. Rautela GS, King KW (1968) Arch Biohcem Biophys 123:589
231. Ray DL (ed) (1959) Marine Boring and Fouling Organsms. University of Washington Press, Seattle
232. Reese ET (1969) Adv Chem Ser 95:26
233. Reese ET (ed) (1963) Advances in enzymic hydrolysis of cellulose and related materials. Pergamon, New York
234. Reese ET (1956) Appl Microbiol 4:39
235. Reese ET (1957) Ind Eng Chem 49:89
236. Reese ET (1965) J Ferment Technol 43:62
237. Reese ET, Gilligan W (1954) J Text Res 24:663
238. Reese ET, Mandels M (1971) In: Bikales NM, Segal L (eds) Cellulose and cellulose derivatives. Series on high polymer, vol 5. Wiley, New York, p 1079
239. Reese ET et al. (1969) Can J Biochem 46:25
240. Reese ET et al. (1957) J Text Res 27:626
241. Reese ET et al. (1950) J Bacteriol 59:485
242. Renkin EM (1954) J Gen Physio 38:225
243. Rogers CJ et al. (1972) Environ Sci Technol 6:715
244. Rosenberg SL (1976) Proc. AIChE 81st National Meeting, Kansas City, Mo.
245. Ross LW, Updegraff DM (1971) Biotechnol Bioeng 13:99
246. Rowland SP (1975) In: Wilke CR (ed) Biotechnol Bioeng Symp, no 5. Interscience, New York, p 183
247. Ryu DDY et al. (1980) Presented at 6th Intern Ferment Symp, London, Ontario, Canada, July 20–25
248. Saarinen P et al. (1959) Acta Agral Fennica 94:41
249. Saeman JF et al. (1952) Ind Eng Chem 44:2848
250. Sarkanen KV et al. (1980) Presented at the AIChE 89th National Meeting, Portland, Oregon, Aug. 17–20
251. Sasaki T et al. (1977) Presented at the Amer. Assoc. Cereal Chem. Ann. meeting, San Francisco, Oct. 23–28
252. Schurz J (1978) In: Ghose TK (ed) Bioconversion of Cellulosic Substances into Energy Chemicals and Microbial Protein Symp. Proc. BERC IIT Delhi, New Delhi, p 37
253. Scott RW et al. (1969) J Forest Prod 19 (4):14
254. Segal L et al. (1959) J Text Res 29:786
255. Selby K (1969) Adv Chem Ser 95:34
256. Selby K (1963) In: Reese ET (ed) Advances in enzymic hydrolysis of cellulose and related materials. Pergamon, New York, p 33
257. Selby K (1968) In: Walters AH, Elphick JJ (eds) Biodeterioration of materials. Elsevier, Amsterdam, p 62
258. Selby K, Maitland CC (1965) Biochem J 94:578
259. Selby K, Maitland CC (1967) ibid. 104:716
260. Simha R (1941) J Appl Phys 12:570
261. Siu RGH (1951) Microbial decomposition of cellulose. Reinhold, New York
262. Shafizadeh F (1977) TAPPI Forest Biology Wood Chemistry Conf., Madison, June 20–22
263. Sharkov VI, Levanova VP (1960) Zh Priklad Khim 33:2563
264. Spano LA et al. (1977) In: Interagency Energy-Environment Research and Development Program Report, EPA-600/7-77/038
265. Stone JE et al. (1969) Adv Chem Ser 95:223
266. Storvick WO, King KW (1960) J Biol Chem 235:301
267. Storvick WO et al. (1963) Biochem J 2:1106

268. Stranks DW (1959) Forest Prod J 9:228
269. Stuck JD (1973) Ph. D. Dissertation, Enzymatic Hydrolysis of Pure and Waste Cellulose. State University of New York at Buffalo
270. Suga K et al. (1975) Biotechnol Bioeng 17:185
271. Suga K et al. (1975) ibid. 17:433
272. Sullivan JT, Hershberger TV (1959) Science 130:1252
273. Suzuki H et al. (1969) Adv Chem Ser 95:60
274. Tanaka M et al. (1979) J Ferment Technol 57:186
275. Tanford C (1961) In: Physical chemistry of macromolecules. New York, p 317
276. Tarkow H, Feist WC (1969) Adv Chem Ser 95:197
277. Tassinari T, Macy C (1977) Biotechnol Bioeng 19:1321
278. Taylor PC (1952) In: Seaman RG, Merrill AM (eds) Machinery and equipment for rubber and plastics, vol 1. Bill Brothers Pub. Corp., New York, p 13
279. Terui G (ed) (1972) Proc IV Int Ferment Symp, Osaka
280. Toda S et al. (1968) J Ferment Technol 46:711
281. Tomita Y et al. (1974) ibid. 46:701
282. Tomita Y et al. (1974) ibid. 52:233
283. Toyama N (1976) In: Gaden EL Jr et al. (eds) Biotechnol Bioeng Symp, no 6. Interscience, New York, p 207
284. Toyama N, Ogawa K (1978) In: Ghose TK (ed) Bioconversioin of cellulosic substances into energy chemicals and microbial Protein Symp Proc. BERC IIT Delhi, New Delhi, p 373
285. Toyama N, Ogawa K (1966) J Ferment Technol 44:741
286. Toyama N, Ogawa K (1976) In: Wilke CR (ed) Biotechnol Bioeng Symp, no 5. Interscience, New York, p 225
287. Toyama N, Ogawa K (1975) In: Bailey M (ed) Symp Enzymatic Hydrolysis Cellulose. SITRA, Helsinki, p 375
288. Tsao GT (1978) J Chin Inst Chem Eng 9:1
289. Tsao GT (1978) Presented at US-ROC Joint Seminar Ferment Eng. Philadelphia, May 30-June 1
290. Tsao GT et al. (1978) In: Perlman D (ed) Ann Rep Ferment Processes, vol 2. Academic Press, New York, p 1
291. Turbak AF et al. (1980) Chemtech, Jan. 1980
292. Tyagi RD, Ghose TK (1978) In: Ghose TK (ed) Bioconversion of cellulosic substances into Energy Chemicals and microbial Protein Symp. Proc. BERC IIT Delhi, New Delhi, p 585
293. Updegraff DM (1971) Biotechnol Bioeng 13:77
294. Van Dyke BH Jr (1972) Ph. D. Dissertation, Enzymatic Hydrolysis of Cellulose: A Kinetic Study. MIT, Cambridge, Mass.
295. Van Soest PJ (1969) Adv Chem Ser 95:267
296. Waiss AC Jr et al. (1972) J Anim Sci 35:109
297. Walker HG et al (1970) Cereal Chem 47:513
298. Walseth CS (1952) TAPPI 35:228
299. Walters AH, Elphick JJ (eds) (1968) Biodeterioration of materials. Elsevier, New York
300. Wardrop AB (1971) In: Sarkanen KV, Ludwig CH (eds) Lignins. Wiley, New York, p 19
301. Warwicker JO et al. (1966) In: Shirley Institute Pamp, no 93
302. Watanabe T (1968a) J Ferment Technol 46:299
303. Watanabe T (1968b) ibid. 46:303
304. Wheatley MA, Moo-Young M (1977) Biotechnol Bioeng 19:219
305. Whitaker DR (1953) Arch Biochim Biophys 43:253
306. Whitaker DR (1954) ibid. 53:439
307. Whitaker DR (1956) Can J Biochim Physiol 34:489
308. Whitaker DR (1956) ibid. 43:102
309. Whitaker DR (1957) ibid. 35:733
310. Whitaker DR (1959) In: Ray DL (ed) Marine boring and fouling organisms. Univ. Washington Press, Seattle, p 301
311. Whitaker DR et al. (1954) Arch Biochim Biophys 49:257
312. Whitaker DR et al. (1963) J Biochem Physiol 41:671
313. Wilke CR (1977) In: Pilot plant studies of the bioconversion of cellulose and production of ethanol. Lawrence Berkeley Lab., Univ of Calif., Berkeley, Ca., LBL 6860, June

314. Wilke CR, Mitra G (1975) In: Wilke CR (ed) Biotechnol Bioeng Symp, no 5. Interscience, New York, p 253
315. Wilke CR (1978) In: Process development studies of the bioconversion of cellulose and production of ethanol. Lawrence Berkeley Lab., Univ. of Calif., Berkeley, Ca. LBL 6881, Feb.
316. Wilke CR (1978) In: Process development of the bioconversion of cellulose and production of ethanol. Lawrence Berkely Lab., Univ. of Calif., Berkeley, Ca., LBL 7880, Sept.
317. Wilke CR, Yang RD (1975) In: Bailey M et al. (eds) Symp Enzym Hydrol Cellulose. SITRA, Helsinki, p 485
318. Wilke CR, Yang RD (1975) In: Timell TE (ed) Proc VIIIth Cellulose Conf. 1. Wood chemicals – a future challenge. Interscience, New York, p 175
319. Wilke CR et al. (1976) In: Gaden EL Jr et al. (eds) Biotechnol Bioeng Symp, no 6. Interscience, New York, p 155
320. Williams MA, Baer S (1964) Feedstuffs 36:48
321. Wilson RK, Pigden WJ (1964) Can J Anim Sci 44:122
322. Wilson PN, Brigstocke T (1977) Proc Biochem 12:17
323. Wood TM (1968) Biochem J 109:217
324. Wood TM (1969) ibid. 115:457
325. Wood TM (1969) Biochim Biophys Acta 192:531
326. Wood TM (1971) Biochem J 121:353
327. Wood TM (1975) In: Wilke CR (ed) Biotechnol Bioeng Symp, no 5. Interscience, New York, p 111
328. Wood TM (1972) In: Terui G (ed) Proc IV Int Ferment Symp. Osaka, p 711
329. Wood TM, McCrae SI (1972) Biochem J 128:1183
330. Wood TM, McRae SI (1977) Carbohydrate Research 57:117
331. Wood TM, McCrae SI (1975) In: Bailey M (ed) Proc Symp Enz Hydrol Cellulose. SITRA, Helsinki, p 231
332. Woodman HE, Evans RE (1947) J Agr Sci 37:202
333. Youatt G (1962) Text Res J 32:158
334. Youatt G, Jermyn MA (1959) In: Ray DL (ed) Marine boring and fouling organisms. University of Washington Press, Seattle, p 397

4 Acid Hydrolysis of Cellulose

Cellulosic materials consist of three major components, namely, cellulose, hemicellulose, and lignin. The two modes of converting the carbohydrate components into their constituent sugars are enzymatic hydrolysis and acid hydrolysis. The former has been reviewed in the preceding chapter [38, 39, 81]. The present chapter covers the latter with the focus on mechanism and kinetics of acid hydrolysis.

4.1 Mechanism of Acid Hydrolysis

Although cellulose possesses excellent strength and good stability, yet it can be degraded by resorting to a variety of chemical and physical processes under certain conditions. The most common manifestation of its deterioration is a decrease in the average degree of polymerization (DP). Usually this deterioration is accompanied by a chemical modification of the cellulose molecule, such as an increase in its reducing power or development of reactive groups along the chain [89].

4.1.1 Course of Hydrolysis Reaction

When cellulose is hydrolyzed in an acidic medium to glucose, the β-1,4-glucosidic bonds of a cellulose chain molecule are split by the addition of water molecules; this addition yields fragments of shorter chain lengths while preserving the basic structure. One of the newly formed end-groups of chain molecules is a potential aldehyde group possessing reducing power [161].

The existence of acid-sensitive linkages in a cellulose molecule has been widely recognized [7, 30, 100, 118, 151], but this concept was not supported by Achwal et al. [2]. The hypothesis that the breakage of β-1,4-glucosidic bonds occurs randomly in acid hydrolysis of cellulose has also been widely accepted [14, 90, 106, 107, 117, 136]. Another hypothesis postulates that the breakage of such bonds is not random [64, 143, 144]; according to Beall and Jorgensen [14], this hypothesis is applicable only if the initial distribution is uniform.

Hydrolysis of cellulose with concentrated acid proceeds through the formation of cellulose acid complexes; this occurs only after the crystalline structure of cellulose is destroyed by its dissolution or swelling in acid [49, 162], i.e.,

Cellulose → acid complex → oligosaccharides → glucose

Examples of acid complexes are:

$$(C_6H_{10}O_5 \cdot 4\,H_2O \cdot H_2SO_4)_n \qquad \text{with sulfuric acid}$$
$$(C_6H_{10}O_5 \cdot 4\,H_2O \cdot HCl)_n \qquad \text{with hydrochloric acid}$$
$$(C_6H_{10}O_5 \cdot 2\,H_2O \cdot H_3PO_4)_n \qquad \text{with phosphoric acid}$$
$$(C_6H_{10}O_5 \cdot H_2O \cdot HNO_3)_n \qquad \text{with nitric acid}$$

Apparently, acid concentration significantly influences the kinetics and course of hydrolysis. In 40% hydrochloric acid, the cellulose is degraded only to oligosaccharides at about 30 °C. The oligosaccharides are hydrolyzed to glucose according to a first-order mechanism only at a higher temperature [48].

Hydrolysis of cellulose with hot dilute acid proceeds through the formation of hydrocellulose to soluble polysaccharides and then to simple sugars, i.e.,

$$\text{native} \xrightarrow{\text{I}} \text{stable cellulose} \xrightarrow{\text{II}} \text{soluble} \xrightarrow{\text{III}} \text{glucose}$$
$$\text{cellulose} \qquad \text{(hydrocellulose)} \qquad \text{polysaccharides}$$

In this serial reaction, the rate-controlling step is hydrolysis of stable hydrocellulose to soluble polysaccharides [161].

The mechanism of acid hydrolysis of cellulose, involving protonation of glucosidic oxygen, has been repeatedly verified [15, 22, 45, 53, 113, 117] and is widely accepted [10, 113, 119, 153]. The course of such hydrolysis is shown in Fig. 4.1. It involves the rapid formation of an intermediate complex between the glucosidic oxygen and a proton and is followed by the slow splitting of glucosidic bonds induced by the addition of the water molecule. The formation of carbonium cation has two possibilities, depending on the primary site of protonation as indicated in Fig. 4.1. If the cyclic O is protonated, the reaction follows path A-2; otherwise, it follows path A-1. More recently, Szejtli [153] has shown that partial protonation of both oxygen atoms, cyclic O and acyclic O, possibly takes place simultaneously, as shown in Fig. 4.2.

Nevell and Upton [113] have assumed that only steps I and IV are reversible (Fig. 4.3) although in principle all four steps can be reversible. Usually, the rate-controlling step is considered to be step II, but under almost water-free condition, step III would become very slow, thus becoming the rate-controlling step. When the two new chain-ends generated from hydrolysis remain in the solid state and are prevented from further hydrolysis, the reverse of step II (i.e., recombination) might occur to a significant extent. The most stable conformation of the carbonium cation generated in step II is perhaps the half-chair, where (C-1), (C-2), (C-5), and O have been brought into one plane by rotations about the (C-2)–(C-3) and (C-4)–(C-5) bonds. Such rotations are more difficult in cellulose than in simple glycosides. This difficulty is caused by two factors; the D-glucose residue attached to (C-4) is long, and adjacent chain molecules are bonded. When water is added to the system, the interfibrillar swelling will occur, thus exposing additional intermediate glucosidic bonds on fibrillar surfaces. The retardation of step III, due to the lack of water, will be diminished, although the reverse of step II will continue to occur. The overall effect, therefore, will be an increase in the extent of hydrolysis of intermediate glucosidic bonds with an increase in the concentration of water.

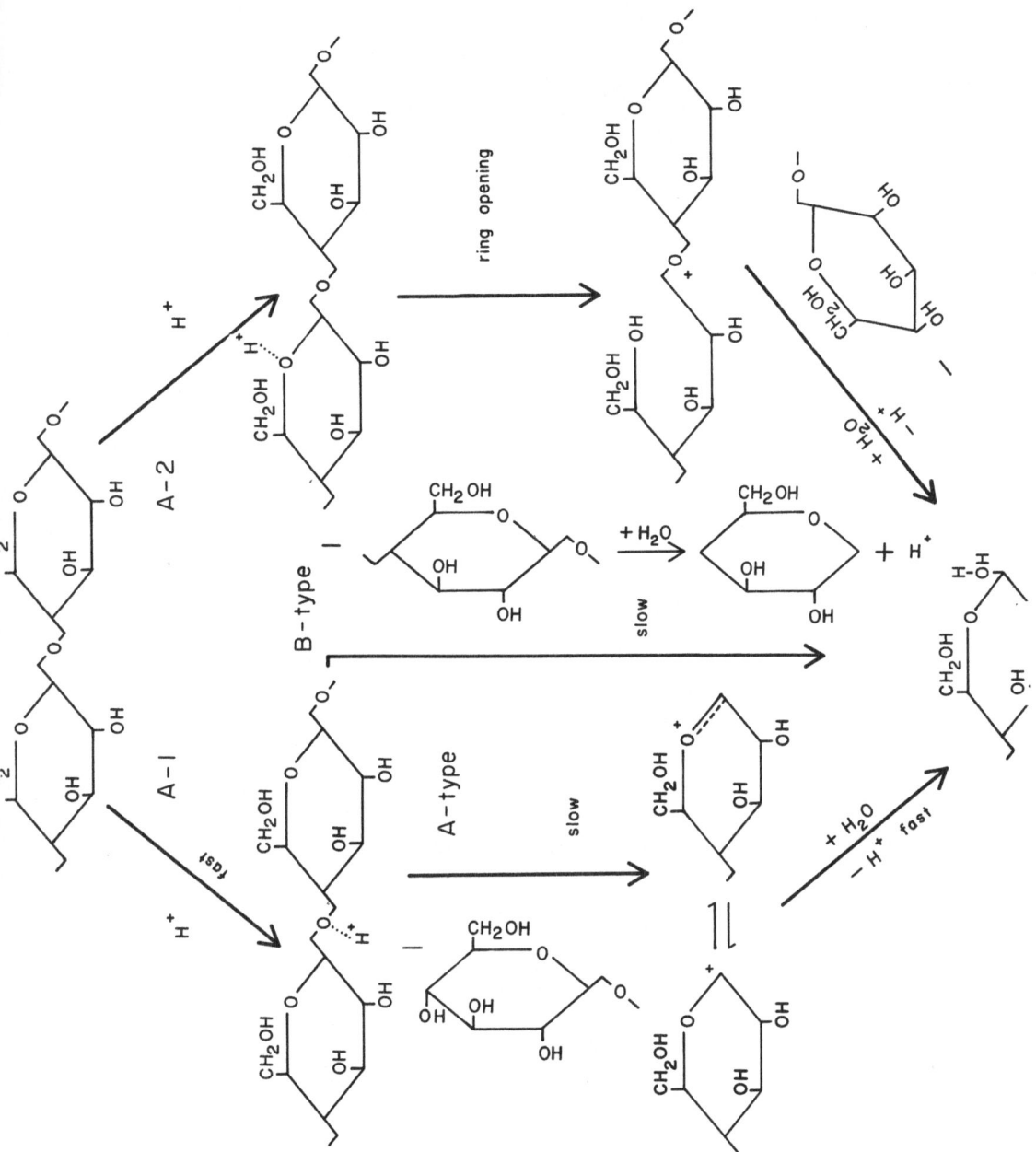

Fig. 4.1. A model of the mechanism of acid hydrolysis of cellulose [22]

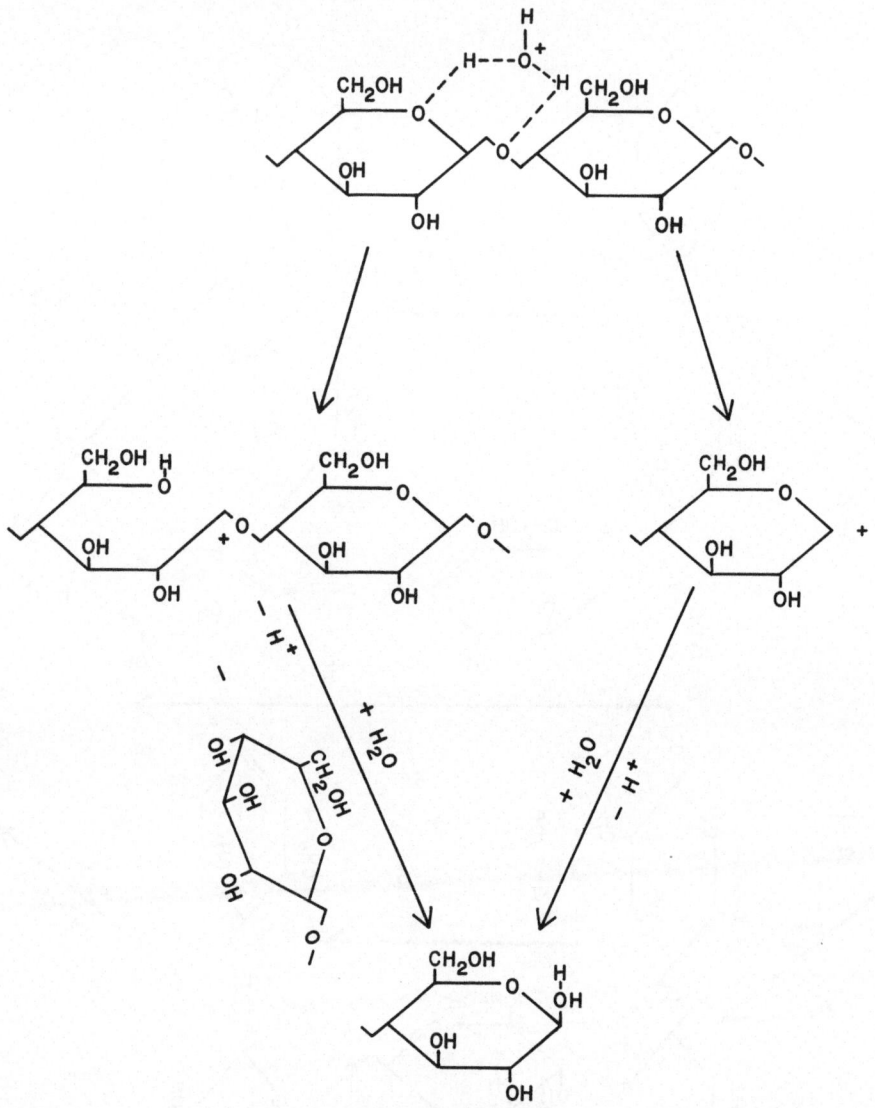

Fig. 4.2. Type A mechanism for acid hydrolysis of cellulose involving protonation of both cyclic oxygen [153]

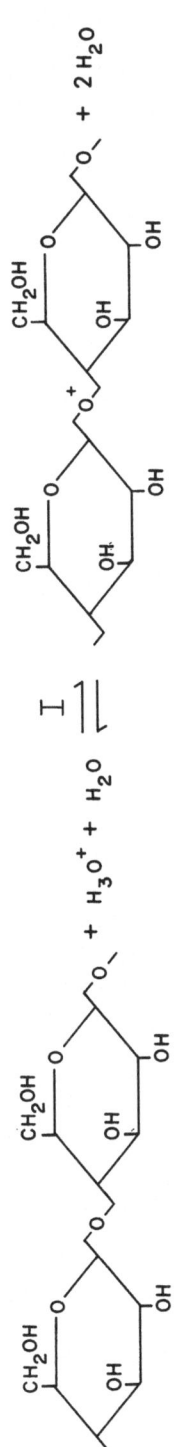

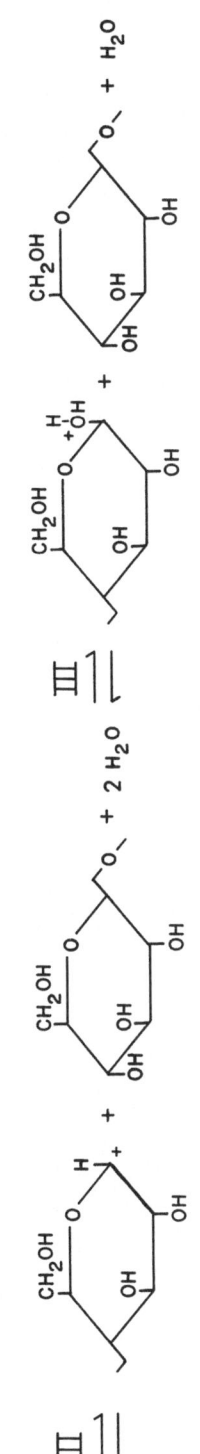

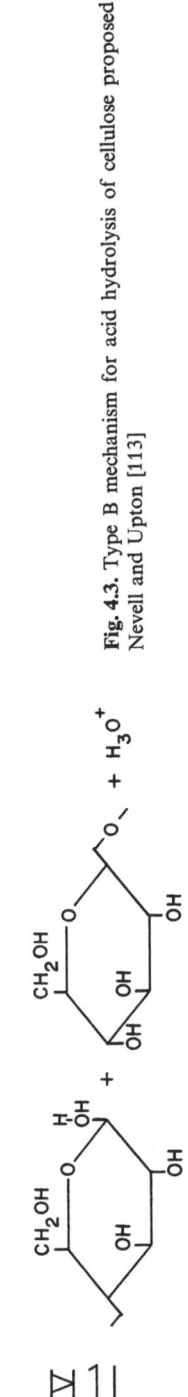

Fig. 4.3. Type B mechanism for acid hydrolysis of cellulose proposed by Nevell and Upton [113]

125

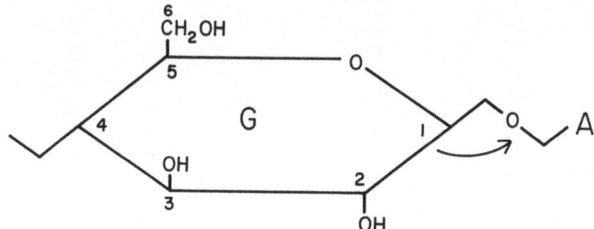

Fig. 4.4. Inductive effects of
neighboring groups [123]

Table 4.1. The effect of electronegativity of the aglycone moiety on the rate of hydrolysis[a] [57, 165]

D-Glucopyranoside	Half-line (min)	Activation energy (kcal/g-mole)	$k \times 10^6$ (min^{-1})
α-Methyl	207,000	34.8	3.35
α-Benzyl	116,000	34.1	5.88
α-Phenyl	3,150	31.1	220
β-Methyl	104,000	33.5	6.66
β-Benzyl	69,700	31.5	9.95
β-Phenyl	11,500	32.2	60.3

[a] Hydrolysis conditions: 0.05 N hydrochloric acid at 60 °C.

According to Bunton et al. [22], Ranby [122], Ranby and Marchessault [123], the rate-controlling step depends on the interaction among neighboring groups, i.e., on the electron transfer during hydrolysis. They proposed that the rate of electron shift is a function of inductive effects from the electron-withdrawing groups in the glycosyl (G) and aglycone (A) (Fig. 4.4); this, in turn, would cause stabilization or activation of the glucosidic bond against acid hydrolysis.

Hydrolysis of a glucopyranoside is increased with increasing electronegativity of the aglycone (Table 4.1) [57, 165]. The aglycone is electronegative in the order of:

$$-CH_3 < -CH_2-C_6H_5 < -C_6H_5.$$

The inductive effects of various glucosyls have not been fully understood. Richards [125] has suggested that the adjacent hydroxyl groups in the (C-2) and (C-3) positions in an acidic medium are present as oxonium groups. This positive charge will inhibit the approach of proton in H_3^+0 to the glucosidic oxygen. Removal of the hydroxyl groups by formation of deoxy sugars is expected to promote these inductive effects. Some experimental data are in good agreement with this hypothesis (see Table 4.2). This approach is supported by the finding that OH(2) and OH(3) are replaced with other electronegative groups, such as chlorine and iodine, thus resulting in the high resistance of the adjacent glucosidic bond to hydrolysis [114].

The availability of hydroxyl groups at (C-2), (C-3), and (C-6) on the D-glucopyranosyl units of hydrocellulose has been studied by Rowland et al. [131, 132, 135, 137]. They have determined that the hydroxyl groups in hydrocellulose

Table 4.2. Rates of hydrolysis of glycosides and deoxyglucosides[a] [63]

Methyl glycoside	$k \times 10^3$ (min^{-1})
α-D-Glucopyranose	2.3
3-Deoxy—D-Glucopyranose	17
α-D-Mannopyranose	4.85
3-Deoxyl—D-Mannopyranose	26.5

[a] Hydrolysis condition: 1 N sulfuric acid at 100 °C.

formed during hydrolysis attain the maximum selective availability at a specific moment. The hydroxyl groups at (C-2) in cotton cellulose are more readily available than those at (C-6) and (C-3) [133]. The experimentally determined relative availabilities of OH(2):OH(3):OH(6) for a highly ordered hydrocellulose II are 1.0:0.26:0.72, compared to estimated values of 1.0:0.0:0.75 for surfaces of elementary fibrils of an idealized perfectly ordered Cellulose II [160]. This has been interpreted to be a reflection of an ordered presentation of hydroxyl groups on the surfaces of crystalline microstructural units [127].

4.1.2 Identification of Mechanism

The course of hydrolysis of cellulose can be classified into A-1 and A-2 types, depending on the site of protonation, as shown in Fig. 4.1 [22]. In the A-2 type, the conjugate acid, which has the proton and positive charge, is transferred to the cyclic O, resulting in ring opening between the cyclic O and (C-1). In the A-1 type, the conjugate acid undergoes heterolysis to form a carbonium cation with the glucosidic ring. The hydrolysis of cellulose can also be type A or B, depending on participation of a water molecule. In type A, a water molecule is not involved, and therefore, the rate-controlling step is unimolecular (Fig. 4.2). According to Hammett and Paul [53], a semilogarithmic plot of the observed rate constant k, against the acidity function $(-H_0)$ is linear thereby supporting the type A mechanism. In the case of type B (Fig. 4.3), the rate-controlling step is bimolecular, and thus, the plot of log k against $(-H_0)$ should not be linear, while a semilogarithmic plot of k against pH should be linear. This criterion has been used for identifying or differentiating mechanisms of hydrolysis of cellulosic materials [109]; nevertheless, it has been shown that departure from linearity is appreciable at higher acid concentrations [84].

Long and Paul [85] have proposed that all A-1 type reactions with uncharged reactants follow the rule according to which the plot of log k against $(-H_0)$ should be linear and that all A-2 type reactions with uncharged reactants follow the rule according to which the plot of log k against pH should be linear. However, Whalley [162, 163] has found that the acid hydrolysis of diethylether proceeds according to an A-2 type mechanism, while the dependence of its rate on H_0 conforms to the A-1

127

type reaction criterion. This criterion, therefore, cannot be used to verify an A-1 type mechanism. Whalley has suggested the following. First, if the overall activation entropy including all the reactions, ΔS^*, is negative, the mechanism is probably of the A-2 type; this probability increases as ΔS^* decreases. Second, if $\Delta S^* > 0$, the mechanism is probably of either the A-1 or A-2 type; the probability of being the A-1 type increases as ΔS^* increases. A large positive ΔS^* for the proton transfer step implies that the validity of this criterion is probably questionable.

4.1.3 Degree of Polymerization

In heterogeneous hydrolysis of cellulose, acid first penetrates into the disordered regions where the splitting of a few glucosidic bonds leads to the rapid decrease in the DP. The subsequent attack is mainly on the crystallites which break down gradually, and the rate of change in the DP decreases gradually. After consumption of the amorphous portion, the DP appears to remain at a constant value, the so-called leveling-off degree of polymerization (LODP).

The LODP depends on the type of cellulose and the mode of hydrolytic pretreatment. Table 4.3 lists the LODP of various types of cellulose samples [13]. LODP is reached when only 2–5 % of the sample has been hydrolyzed [13, 32, 111]. The LODP is considered as the average length of crystallites in the cellulose sample [12, 13, 108, 115, 130].

It has been shown [16, 62, 66, 108] that the values of the average length of crystallites determined by an electron microscope are in agreement with those derived from LODP measurements. It has also been shown [91] that the LODP may be an artifact introduced by washing the residue from hydrolysis with water, thus removing the low molecular weight components by dissolution in water. A method of rapidly measuring the cellulose viscosity for the purpose of calculating DP has been proposed by Alexander and Mitchell [4]. Nelson and Tripp [111] have forwarded a graphical method for determining the value of DP.

Table 4.3. LODP of cellulosic materials [a] [12]

Type of cellulose	Average LODP
Natural fibers	
Ramie, hemp	350 ~ 300
Cotton, purified	250 ~ 200
Unbleached sulphite wood-pulp	400 ~ 250
Bleached sulphite pulp	280 ~ 200
Mercerized cellulose (18 % NaOH *at* 20 °C, 2 h)	90 ~ 70
Vibratory milled wood cellulose	100 ~ 80
Regenerated fibers	
Fortisan, fiber G	60 ~ 40
Textile yarns	50 ~ 30
Tire yarns	30 ~ 15

[a] Hydrolysis conditions: 2.5 N HCl at 105 °C for 15 min.

128

Many investigators have discussed the factors influencing the DP. Battista and Coppick [13] demonstrated the strong dependence of the DP of residues from hydrolysis or residual cellulose on the conditions of solution and regeneration. In an early work, Davidson [33] observed that the preswelling of cotton cellulose with sodium hydroxide solution followed by acid hydrolysis increased the extent of hydrolytic degradation. Jorgensen [65, 66] further exhibited that the DP of residues depended on the extent of preswelling of native cellulosic materials by sodium hydroxide solutions. Millett et al. [103] and Daruwalla and Shet [32] used very severe hydrolytic conditions; their results indicate clearly that the DP of residues is related to the severity of hydrolytic conditions. Daruwalla and Shet [32] carried out heterogeneous hydrolysis of Sudance cotton with HCl of 2, 4, and 6 N at 30°, 60°, and 90 °C. Their results apparently indicate that the rate of decrease in the average DP is enhanced by elevation of both the acid concentration and temperature. Their results also indicate that the LODP is influenced neither by acid concentration nor by temperature of hydrolysis, and that the values obtained are between 80 and 90. These values are in fairly good agreement with those obtained by other investigators. For example, Jorgenson [67] obtained a value of 60; Millett et al. [103] 88; Battista et al. [12] 70–90. These results correspond to crystallite lengths of approximately 400–450 Å, which are close to the particle length obtained by Immergut and Rånby [62] for hydrolyzed residue from Empire cotton. In contrast, the value obtained by Nelson [109] 150, and that by Davidson [33] 190, are higher than those obtained by Daruwalla and Shet [32]. The main reason is attributed to the choice of the constant for converting intrinsic viscosity to the DP value.

The variations of average DP of the mercerized cotton yarns may be divided into the following two cases. First, the mercerization-stretching under mild tension brings about a decrease in the average DP. For example, slack mercerization followed by stretching of the cotton yarns up to 96% of the original length in the mercerizing solution reduces the aberage DP, as shown in Table 4.4 (see rows II, III, and IV) [56]. The mercerization-stretching, however, brings about an increase in the average DP under high tension [56, 97]. Stretching of cotton yarns to 100% or 103% of the original length resulted in a noticeable increase in the average DP (see

Table 4.4. Effect of mercerization and stretching on DP of cotton yarn [56]

Substrate [a]	DP after a time of acid hydrolysis (min) [b]					
	0	5	15	30	60	90
I	2412	2052	2122	2079	2094	2049
II	2342	1948	1989	1950	1985	1930
III	2362	2183	2100	2118	1954	2005
IV	2396	2195	2088	2066	2094	1885
V	2504	2164	2106	2178	2079	1890
VI	2504	2210	2245	2200	1988	1962

[a] I. Scoured yarn; II. mercerization as nearly slack as possible on the machine, followed by partial restretching to 90% of the original length; III. ibid, but to 94%; IV. ibid, but to 96%; V. ibid, but to 100%; VI. ibid, but to 103%.
[b] Hydrolysis condition: 0.03 N. hydrocholoric acid at 40 °C.

rows V and VI in Table 4.4). This is due to removal of low molecular weight components during the mercerization.

When scoured cotton yarns (row I in Table 4.4) and mercerized cotton yarns (rows II–VI in Table 4.4) are treated with HCl, the DP decreases with time, depending on the conditions employed [56]. Table 4.4 also shows that there is a remarkable decrease in the average DP, particularly in the initial stage of hydrolysis; subsequently, the average DP decreases gradually and slowly. The decrease in the average DP of mercerized cottons is much greater than that of scoured cotton. In other words, mercerized cotton yarns undergo a greater extent of degradation by acid hydrolysis than scoured cotton yarns [56].

It has been proved [56, 156] that mercerization transforms the molecular structure of cellulose fibers in such a way that mercerized cellulose fibers show higher strength than the corresponding unmercerized fibers. For example, unmercerized cotton cord loses 25–50% of its strength during hydrolysis, when the fraction of bonds broken ranges from 0.064 to 0.1%, whereas mercerized yarns lose 25–40% of their strength when the fraction of bonds broken ranges from 0.15 to 0.25% [156].

Pretreatment of cotton fibers with ethylenediamine (EDA) and caustic soda (NaOH) has been shown to increase the reactivity of fibers towards acid hydrolysis. At the end of rapid acid hydrolysis, Modi et al. [104] reported viscosity average DP values of 616 for untreated cotton fibers, 247 for cotton fibers treated with caustic soda, 355 for cotton fibers treated with ethylenediamine (EDA), and 448 for cotton fibers treated with zinc chloride. Although pretreatment with caustic soda, ethylenediamine, or zinc chloride has been observed to promote the degradation of cotton fibers, the pretreatment with caustic soda and zinc chloride provide the largest and the smallest increase in reactivity, respectively.

Oxidation of cellulose and hydrocellulose prior to hydrolysis or during progressive hydrolysis reduces the DP of partially hydrolyzed residues [1, 21, 130, 134]. The action of oxidation decreases the content of aldehyde groups in the molecular chains of hydrocellulose and increases the content of carboxyl groups [1, 34, 35, 99]. The presence of carboxyl groups in the amorphous regions during hydrolysis prevents recrystallization that would generally be expected to increase the DP slightly [129, 130]. It has been shown that oxidation induces some changes in the fine structure of cellulose and that these changes may influence its properties, such as moisture regain capacity, density, and DP [94, 96].

The average DP of cellulose can be determined routinely by viscosity measurement [21]. However, characterizing these polymers in terms of their molecular weight distributions (MWD) or degree of polymerization distributions (DPD) is more useful in production control and end-use performance evaluation. More recently, the DPD's of various types of celluloses and their derivatives during either acid hydrolysis or enzyme degradation have been studied by gel-permeation chromatography (GPC) [5, 126, 139, 145]. On the basis of the folding chain model, Chang et al. [27] have explained that the rapid initial decrease in the average DP observed in their experiments is due to the splitting of a few glucosidic bonds in the disordered regions; subsequent attack is mainly on the crystallites, which break down gradually into molecular segments, including species with an average DP of 8. Further degradation of these hydrolysis products is prevented. Barth and Regnier

[9] used GPC with a hydrophilically coated silica as packing and a high ionic strength buffer as the mobile phase in order to determine the DPD of modified cellulose. They have shown that various water-soluble celluloses may be analyzed within 15 min by using this system.

For many years, GPC analysis of cellulosic materials was carried out by first making them soluble in the solvents. This process is time consuming, and furthermore, it degrades samples. Bao et al. [8] developed a new approach, using sepharose CL-6B as packing, 0.5 N sodium hydroxide as the effluent and cadoxen as the cellulose solvent. This technique was employed to study the change in cellulose structure after solvent pretreatment. Bose and Tsao [19] observed that the DP of wood pulp pretreated with cadoxen for 4 h was 936 while that of the untreated sample was 929. These results indicate that cadoxen pretreatment causes insignificant degradation of a cellulose sample. On the other hand, the DP of the same sample, when pretreated with 70% H_2SO_4 for 1 h, dropped to 878. This indicates that acid pretreatment causes some degradation of cellulose, but its magnitude is small.

4.1.4 Recrystallization

Recrystallization is the process of crystal growth in the fine structure of cellulose during either acid or enzymatic hydrolysis. This implies that the crystallinity index of cellulose increases as hydrolysis proceeds.

A controversy exists between the hypotheses of amorphous structure and completely crystalline protofibers [83]; however, the recrystallization during heterogeneous acid hydrolysis is commonly recognized. Hermans and Weidinger [59] used the X-ray technique to measure the crystallinity of cellulose during heterogeneous acid hydrolysis. They have found that during hydrolysis with boiling 2.5 N sulfuric acid solution, the crystalline fraction in viscose rayon increases rapidly from 39 to 49% within 0.5 h, corresponding to an increase of approximately 10% of the original cellulose and that the crystalline fraction is unchanged in quantity after completion of the recrystallization. Achwal et al. [1] observed a significant increase in crystallinity of all the cellulosic materials examined. Battista [11] studied the recrystallization of cellulose by using both mild (0.5 N hydrochloric acid at 5, 18, and 40 °C) and drastic (2.5 N and 5.0 N hydrochloric acid at boiling point) conditions for hydrolysis; he found that the effect of recrystallization on hydrolysis is more pronounced for regenerated cellulose than for native cellulose. Sharples [148] showed that, on hydrolysis of Egyptian cotton with 0.1 N sulfuric acid at 50 °C, the amount of accessible material at the greatest extent of degradation is only 5%, compared with 10.8% found in ungraded cotton; furthermore, recrystallization takes place at the early stage of hydrolysis, starting approximately when one-eighth of the intercrystalline chain segments is broken. Even though the crystallinity seems to be unchanged in the subsequent stage, a slight increase in density is detected [31]. Shinouda and Moteleb [152] have observed that the crystallinity index increases approximately 30% when viscose fibers are hydrolyzed in an acidic medium. Rozmarin [138] has proposed a mathematical expression to describe the effects of acid concentration, temperature, reaction time and liquor-to-cellulose ratio on the crystallinity index for hydrolysis of cotton; the calculated results agree well with the experimental values.

Heterogeneous hydrolyses of ramie fiber [59], wood pulp [58], slash pine sulfated pulp [91], and gum sulfated pulp [91] do not appear to generate crystallites. According to Sharples [149], although the amorphous content in any of these cellulosic materials is initially small, it is reduced further by recrystallization. The fraction of amorphous portion removed during the early stage of hydrolysis is only approximately 5%, and thus, the amorphous content in the remaining crystalline material is still approximately 30%. The overall change in crystallinity is, therefore, barely detectable by the X-ray technique.

The effect of swelling treatments on the crystallinity of cellulose has been studied by numerous investigators [20, 40, 54, 60, 84, 101, 142, 153]. The crystallinity of rayons increases from 40 to 50% after treatment with 18% NaOH [60]. Oxygen-containing polar solvents may cause a reconversion of the crystal structure to cellulose I and also promote recrystallization [146]. The crystallinity index of Solka Floc measured by X-ray diffraction is 74.2; this value increases to 77.4 after swelling with water and then treatment with methanol and benzene [40]. Cellulose I lattice is highly resistant to alkaline attack; it is capable, however, of recrystallization on subjecting the swollen cellulose to tension [87]. The amount of recrystallized fraction is proportional to the residual cellulose I in the slack-mercerized samples; the presence of small crystallites of cellulose I is a prerequisite for further growth of the cellulose I lattice upon stretching.

In contrast to NaOH and the oxygen-containing polar solvents mentioned above, the treatment of cellulose with certain swelling agents brings about a decrystallization and an increase in reactivity. It has been demonstrated [18, 19, 80] recently that several important structural features of cellulosic materials are affected significantly by both chemical and physical pretreatments. The chemical and physical pretreatments of cellulose prior to acid-catalyzed hydrolysis have also been investigated widely [6, 37, 47, 70, 71, 84, 86, 104, 112, 124, 164]. These works have been reviewed by Millet et al. [102], Toyama et al. [155], Halliwell [52], Lin et al. [82], Chang et al. [24], and Fan et al. [39]. In fact, recrystallization essentially occurs when partial or total amorphous cellulose is exposed to water, even by wetting in atmospheric moisture [23, 159]. When Solka Floc with a crystallinity index of 74.2 was placed in air saturated with water vapor at 24 °C, its crystallinity index rose to 81.8 eventually; this might be due to the moisture helping platicize longer segments of amorphous cellulose chains and rendering them to align under the action of hydrogen bonds [16].

It should be pointed out that two different mechanisms of recrystallization exist. In mild hydrolysis, recrystallization is attributable to the extension of the crystalline regions as a result of the splitting of 1,4-glucosidic bonds in cellulose molecules, thus causing the aggregation of short and long chain segments [135]. Under the drastic conditions of hydrolysis, the splitting of 1,4-glucosidic bonds is accompanied by the formation of short chains. Therefore, the intercrystalline chain segments are removed selectively, and the loose chain ends undergo crystallization relatively freely. As the increasing number of chains continue to be cut, the increase in crystallinity becomes more prominent, rendering the crystallinity of hydrolyzed cellulose to be higher than that of unhydrolyzed cellulose [1].

4.2 Kinetics of Acid Hydrolysis of Cellulose

In this section, kinetic models for acid hydrolysis of cellulose are reviewed.

4.2.1 Homogeneous Kinetics

A. Deterministic model

As mentioned previously, the acid hydrolysis of cellulose in a homogeneous phase, in which the rate-determining step is the splitting of glucosidic bonds in stable hydrocellulose, follows the first-order reaction kinetics; thus, it can be written as [149]

$$\frac{dn}{dt} = -kn, \tag{4.1}$$

where k is the rate constant for splitting of glucosidic bonds, t hydrolysis time, and n the total number of bonds available in all the cellulose molecules in the system. The relationship between n and P_n, which is the number-average degree of polymerization, may be expressed as

$$n = N - \frac{N}{P_n}, \tag{4.2}$$

where N is the number of repeating monomer units in the system. Integrating Eq. (4.1) from time 0, when the number-average degree of polymerization is P_n^0, to time t, when the number-average degree of polymerization is P_n results in,

$$k = \frac{1}{t} \ln \left[\left(1 - \frac{1}{P_n^0} \right) \bigg/ \left(1 - \frac{1}{P_n} \right) \right]. \tag{4.3}$$

If P_n^0 and P_n are large, the extent of hydrolysis is small. This represents the situation at the very early stage of hydrolysis, and under this situation, Eq. (4.3) reduces approximately to

$$k = \frac{1}{t} \left(\frac{1}{P_n} - \frac{1}{P_n^0} \right). \tag{4.4}$$

This expression has also been obtained by Schulz and Lohmann [144]. Equation (4.4) indicates that the plot of $(1/P_n)$ against t should yield a straight line, and that the rate of hydrolysis may be determined by measuring the mean molecular weight of DP during hydrolysis.

Moiseev et al. [105] have established a kinetic equation from spectropolarimetric data on homogeneous acid hydrolysis of O-methylcellulose; it is

$$k = \frac{1}{t} \ln \left[(\theta_\infty - \theta_0)/(\theta_\infty - \theta) \right], \tag{4.5}$$

where θ_0, θ, and θ_∞ are rotation angles of the solution at the onset of the experiment, at time t, and at the end of the experiment, respectively. This equation indicates that the plot of $\ln(\theta_\infty - \theta)$ against t should be linear.

Chang [26, 29] has found that his results cannot be modeled by Eq. (4.4). According to him, the mathematical expression based on the folding-chain model agrees well with the hydrolysis results for cotton linters cellulose in 1 N hydrochloric acid at 80 °C [26, 29], and those for HWM-rayon in 1 N hydrochloric acid at 50 °C [25]. This model is as follows:

$$k = \frac{1}{t} \ln \left[\left(1 - \frac{F_L}{P_n^0} \right) \Big/ \left(1 - \frac{F_L}{P_n} \right) \right],$$

(4.6)

where k is the rate constant for cutting the weak β_L-bonds at folds, F_L the fold length which can be estimated approximately by the expression,

$$F_L = \frac{P_n^0}{m},$$

(4.7)

where m is the number of segments per molecule, which has been found to be 7 [24, 25].

Saeman [140] has proposed a homogeneous kinetic model for hydrolysis of wood cellulose. The model visualizes insoluble cellulose as being present in the form of the equivalent amount of dissolved glucose. Furthermore, the model assumes that the hydrolytic degradation proceeds in two successive first order irreversible reactions; in the first reaction, cellulose is decomposed to glucose that, in turn, is decomposed in the second reaction as illustrated below.

$$\text{cellulose} \xrightarrow{k_1} \text{glucose} \xrightarrow{k_2} \text{decomposed glucose}$$
$$\qquad A \qquad\qquad\quad B \qquad\qquad\qquad\quad C$$

The resultant kinetic equations are written as

$$\frac{dC_A}{dt} = -k_1 C_A,$$

(4.8)

$$\frac{dC_B}{dt} = k_1 C_A - k_2 C_B,$$

(4.9)

$$\frac{dC_C}{dt} = k_2 C_B,$$

(4.10)

$$k_1 = K_1 (C_\sigma)^m \exp(-E_1/RT),$$

(4.11)

$$k_2 = K_2 (C_\sigma)^n \exp(-E_2/RT),$$

(4.12)

where C_A, C_B, and C_C are the concentrations of cellulose, glucose, and decomposed glucose, respectively, C_σ the acid concentration, E_1 the activation energy of the first

134

step, E_2 that of the second step, R the gas constant, and T the absolute temperature; K_1, K_2, m, and n are the empirical constants. It has been shown that this kinetic model is applicable to many cellulosic materials, but the parameter values are different for various materials. The parameter values determined by Kobayashi et al. [74], Fagen et al. [36], Vaux [158], Thompson and Grethlein [154], Planes [121], and McKibbin [92] are summarized in Table 4.5

Glucose is the sugar of primary interest in cellulose hydrolysis. Its optimal concentration at a given temperature may be estimated as follows [51].

(i) Combining Eqs. (4.8) and (4.9) yields the following expression of dependence of the glucose concentration, C_B, on the reaction time, t;

$$C_B = C_A^0 \left[\frac{k_1}{k_1 - k_2} \right] [\exp(-k_2 t) - \exp(-k_1 t)] + C_B^0 \exp(-k_2 t), \quad (4.13)$$

where C_B^0 is the initial glucose concentration.

(ii) Equating the time derivative of Eq. (4.13) to zero and solving the resultant algebraic expression gives the optimum reaction time, t_{opt}, as

$$t_{opt} = \frac{1}{k_2 (k_r - 1)} \ln \left(\frac{k_r}{H} \right), \quad (4.14)$$

where

$$k_r = \frac{k_1}{k_2} = \frac{K_1}{K_2} (C_a)^{(m-n)} \exp(E_2 - E_1)/RT, \quad (4.15)$$

$$H = 1 + \frac{C_B^0}{C_A^0} \left(\frac{k_r - 1}{k_r} \right). \quad (4.16)$$

(iii) By substituting Eq. (4.14) into Eq. (4.13), an expression for the maximum glucose concentration is obtained as

$$(C_B)_{max} = C_A^0 \left(\frac{H}{k_r} \right) \frac{1}{k_r - 1}. \quad (4.17)$$

If $C_B^0 = 0$ or $H = 1$, Eq. (4.17) reduces to

$$(C_B)_{max} = C_A^0 \frac{1}{k_r (k_r - 1)}. \quad (4.18)$$

Equation (4.17) or (4.18) indicates that the optimum glucose concentration is a function solely of the ratio of k_1 and k_2; this ratio,

$$k_r = k_1/k_2,$$

is a function of the reaction temperature and acid concentration. Figure 4.5 gives some profiles predicted by Eq. (4.13) [36]. We see that the maximum glucose yield

Table 4.5. Kinetic parameters in the Saeman model for acid hydrolysis of various substrates

Parameters	K_1 (min^{-1})	K_2	E_1 (cal/mol)	E_2	m	n	Temp. (°C) acid conc. (wt-%)	Ref.
Douglas fir	1.73×10^{19}	2.38×10^{14}	42,900	32,800	1.34	1.02	170–190°C 0.4–1.0% H_2SO_4	[140]
Glucose	–	1.85×10^{14}	–	32,690	–	1.0	160–260°C	[92]
Kraft paper	28×10^{19}	4.9×10^{14}	45,100	32,800	1.78	0.55	180–230°C 0.2–1.0% H_2SO_4	[36]
Newsprint	28×10^{19}	4.8×10^{14}	45,100	32,800	NR	NR	200–240°C 1% H_2SO_4	[158]
Solka-floc	1.22×10^{19}	3.79×10^{4}	42,500	32,700	1.16	0.69	180–240°C	[154]
Canebagasse	1.15×10^{21}	–	36,300	–	0.64	–	40–80°C 15–50% H_2SO_4	[121]
Cellulose	1.57×10^{14}	–	34,000	–	1.42	–	100–130°C 5–40% H_2SO_4	[74]

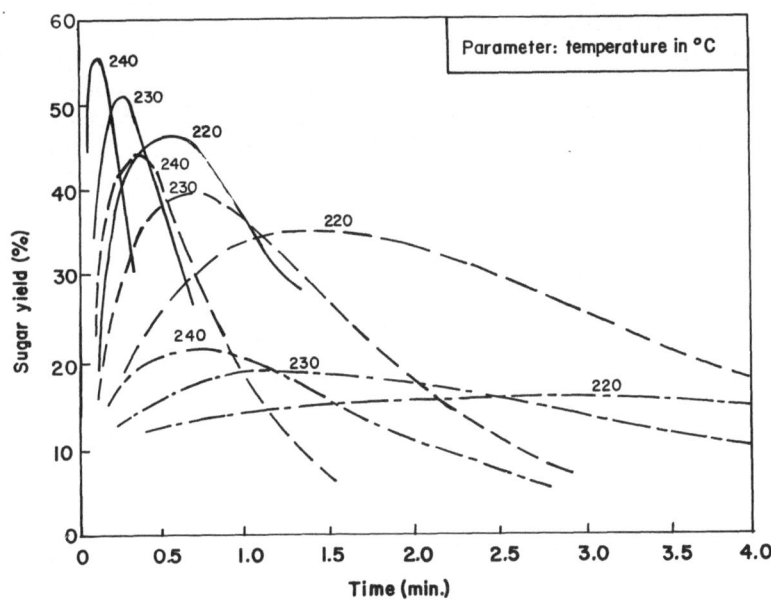

Fig. 4.5. Dependence of the predicted sugar yield on potential glucose, and hydrolysis time; —, 1% acid; ---, 0.5% acid; —·—, 0.2% acid [36, 51]

may be obtained at the highest temperature (pressure) and acid concentration within a short reaction time. It is rather difficult, however, to achieve a complete saccharification of cellulose due to the small difference between the E_1 and E_2. Recent investigations have shown that this kinetic model favors the use of a plug flow reactor [50, 51, 128, 154].

Harris and Kline [55], Kobayashi [73], and Okamura [116] studied the kinetics for hydrolysis of cellulose in the presence of excessive acid. Freudenberg and Kuhn [46], Freudenberg and Blomquist [45], and Klages [68, 69] proposed kinetic equations for hydrolysis of oligosaccharides. Klages found that Freudenberg and Kuhn's equation could not be applied to cellotriose and cellotetraose when hydrolysis was carried out in 51% sulfuric acid at 18°–30 °C. Klages was successful in applying his kinetic expression to octamethylcellobiose and hendecamethylcellotriose when hydrolysis was carried out with 0.1 N hydrochloric acid at 98 °C and to the very early stages of hydrolysis of cellotriose and that of cellotetraose. It seems that strong sulfuric acid is somewhat "anomalous" in the action on cellobiose and cellulose at low temperatures (18–30 °c) [95].

It must be noted that the mode and rate of hydrolysis of cellulose by a small amount of sulfuric acid differ from those by a large amount of the same acid. In the former situation, hydrolysis proceeds as a successive first-order reversible reaction of the A $\rightleftarrows$ B $\rightleftarrows$ C type with respect to cellulose, while in the latter situation, it proceeds as a consecutive first-order reaction as mentioned previously. Kinetic models for the former have been established by Goto et al. [49], Kobayashi et al. [72], and Sakai [141].

137

B. Stochastic Model

Kinetics of depolymerization of macromolecules (100 or more monomer units) seems to be susceptible to a statistical or stochastic treatment [107]. Kuhn [76] appears to be the first to contribute to the development of statistical theory of cellulose depolymerization. In deriving his model, Kuhn has made assumptions that the initial chains are infinitely long compared to the degraded products; all chains are equally reactive; all initial molecules have a uniform size; and the accessibility of a bond to reaction is independent of its position in the chain and length of its parent chain. His model has been frequently verified experimentally [42–44, 46].

A model based on the assumption of infinite initial chain lengths fails to predict the experimental results on the extent of cellulose degradation; this extent of cellulose degradation manifests itself through the variation in molecular weight distribution. Investigation, therefore, of depolymerization based on the finite and uniform initial chain length has received considerable attention. Montroll [106] has statistically treated this case; he has also proposed a relationship between the depolymerization rate constant k during hydrolytic cleavage of a linear polymer and the average DP of the degraded products for the case with initial molecules of finite and nonuniform sizes. In deriving this relationship, the initial distribution of molecular weights has been assumed to be normal or Gaussian [107].

For the most general case, i.e., the depolymerization of chains with finite but nonuniform lengths with different probabilities of breaking different bonds, Beall and Jorgensen [14] have developed an approach that relates the statistical moments of distribution of the degraded polymer to the corresponding moments of the initial material. It has been shown that their model agrees well with the experimental data obtained by hydrolyzing cellulose contained in two types of cotton [89].

4.2.2 Heterogeneous Kinetics

A. Empirical Model

Since the mechanism of heterogeneous hydrolysis of cellulose is yet to be clearly deciphered, many investigators in this field have been satisfied with empirical kinetic expressions from their experimental observations.

Birtwell et al. [17] have proposed an equation based on the measurement of end-group, i.e., aldehyde group; this yields the so-called copper number, N, which is directly proportional to the hydrolysis time, t, i.e.,

$$N = ct^n, \qquad (4.19)$$

where c and n are empirical constants. A similar equation has been proposed by Yorston [166] from the-loss-in-weight measurements during hydrolysis, it is

$$X = ct^n. \qquad (4.20)$$

This expression indicates that the percentage loss-in-weight, X, is directly proportional to the time of hydrolysis; c and n are empirical constants. These equations, however, have been found to be valid only in the early stage of hydrolysis.

Nelson and Conrad [110] and Phillip et al. [120] have derived the following equation from the loss-in-weight measurements during hydrolysis.

$$W = \alpha_0 \exp(-r_a t) + (1 - \alpha_0) \exp(-r_c t), \qquad (4.21)$$

where W is the weight fraction of unhydrolyzed cellulose, α_0 is the initial fraction of amorphous cellulose, and thus, $(1 - \alpha_0)$ is the initial fraction of crystalline cellulose, and r_a and r_c are the rates of hydrolyses of amorphous cellulose and crystalline cellulose, respectively.

Meller [93] also employed the loss-in-weight technique and discussed the kinetics of removal of the "difficult-to-access fraction". He has postulated that the temperature coefficient of the rate-controlling reaction in the removal of the "difficult-to-access fraction" from the fibers by 8 % hydrochloric acid at 80–100 °C would correspond to an apparent activation energy of 28–29 kcal/g-mole. The following set of five·equations has been derived by him [96];

$$x = x_a + x_c, \qquad (4.22)$$

where x, x_a, and x_c are, respectively, the fractional loss-in-weights of the "total", "easy-to-access fraction", and "difficult-to-access fraction" measured at any time. The rate of removal of the "difficult-to-access fraction" may be expressed as the zero-order behavior, i.e.,

$$\frac{dx_c}{dt} = k_c, \qquad (4.23)$$

where k_c is the rate constant. The rate of removal of the "easy-to-access fraction" can be divided into three stages, i.e.,

$$\frac{dx'_a}{dt} = k'_a \qquad (4.24)$$

for the initial stage,

$$\frac{dx''_a}{dt} = \frac{k''_a}{(1 + bx''_a)} \qquad (4.25)$$

for the intermediate stage, and

$$\frac{dx'''_a}{dt} = k'''_a(\alpha - x'''_a) \qquad (4.26)$$

for the final stage. In these three equations, k'_a, k''_a, and k'''_a are the rate constants in the initial, intermediate, and final stages, respectively; x'_a, x''_a, and x'''_a are the

139

fractional loss-in-weights at various stages; and α is the fraction of the "easy-to-access fraction". From this set of five equations, we see that: (i) in the early stage, the removal of the readily hydrolyzable components from cellulose is directly proportional to time, and the rate of removal is constant and may be expressed by the zero-order kinetic equation; (ii) in the intermediate stage, the rate of removal of the readily hydrolyzable components is decelerated appreciably due to retardation by the rate-controlling process and is represented by a retarded rate equation; and (iii) in the final stage, the rate of removal of the easily hydrolyzable components is expressed as the first-order kinetic equation. The plot of (dx_a'''/dt) against (x_a''') yields curves terminating in straight lines. These equations have been experimentally verified [96].

Chang et al. [28] have proposed a kinetic model of acid hydrolysis of cellulose, which visualizes that the random attachment occurs on the surface of crystalline cellulose, and the dissolution of resultant small oligomers into the reaction medium causes the loss-in-weight. The model is expressed as

$$W_t = W_0 \left(\frac{P_n' - \alpha}{P_n'} \right) \exp\left(-\alpha\, kt \right), \tag{4.27}$$

where W_0 is the original weight of the sample, P_n' the average length of the crystallite, and α the number of monomer units in the largest soluble oligomer; for hydrolysis in $6\,\mathrm{N\,HCl}$, α can be as high as 12.

Sharples [148] has derived the following equation for the heterogeneous hydrolysis of cellulose.

$$\ln\left(1 - \frac{1}{P_n^0} \right) - \ln\left(1 - \frac{1}{P_n^0} + \frac{1}{\alpha\, P_n^0} - \frac{1}{\alpha\, P_n} \right) = kt. \tag{4.28}$$

When the extent of degradation is low, this expression reduces approximately to

$$\frac{1}{P_n} - \frac{1}{P_n^0} = k\,\alpha\, t. \tag{4.29}$$

The values of P_n, determined by the osmotic method, are not sufficiently accurate in the high molecular weight range while the weight-average degree of polymerization, P_w can be accurately measured by the viscosity method; the ratio P_w/P_n is equal to 2 over a wide extent of degradation range. Equation (4.29), therefore, can be expressed in terms of P_w as follows:

$$\frac{2}{P_w} - \frac{2}{P_w^0} = k\,\alpha\, t. \tag{4.30}$$

This equation shows that the plot of $(2/P_w)$ against time t should be linear. The results from heterogeneous acid hydrolyses of Egyptian cotton and bacterial cellulose in $0.1\,\mathrm{N}$ sulfuric acid at $50\,^\circ\mathrm{C}$ [148] have verified the validity of this equation. Both equations, (4.28) and (4.30), however, have been successfully applied to situations when the extent of degradation is relatively low. When the extent of degradation is high, especially for hydrolysis in dilute acid, significant deviation from linearity has been observed [61, 109]. This deviation may be attributed to the fine structure of cellulose becoming less accessible due to short chain segments generated by the splitting of chains and becoming oriented to form more ordered regions so that fewer glucosidic bonds are accessible to the hydrolysis action [32].

To reduce the deviation from linearity over a wide range of DP, Sharples [148] has suggested the following equations from examination of the distribution of chain scission at various degrees of degradation;

$$P_n = \frac{P_n^0}{1 + YX} \tag{4.31}$$

for cases with all equally effective cuts, and

$$P_n = \frac{P_n^0}{(1 - e^{-X})Y + 1} \tag{4.32}$$

for cases with effective first cuts only. In these equations, X is the average number of cuts per crystalline segment, and Y is the average number of intercrystalline segments present in the undegraded chain molecule evaluated from the crystalline size (number-average DP) and from the accessibility of the unhydrolyzed structure.

Nelson [109] has also proposed an empirical approach for heterogeneous acid hydrolysis. He obtained experimental data by hydrolyzing cotton linters with 6 N hydrochloric acid at 90° and 100 °C, crystallized cotton with 6 N acid at 80° and 100 °C, and cotton linters and viscose rayon with 0.01, 0.1, 1.0, and 2.5 N hydrochloric acid at 80° and 100 °C. He has found that the data can be fitted well with the following hyperbolic equation;

$$\frac{1}{kt} = \frac{1}{P_n - P_n^0} - \frac{1}{P_n' - P_n^0}, \tag{4.33}$$

where P_n' approximates the LODP. From the resultant values of k, Nelson has evaluated the activation energies for the accessible fractions of cotton linters and viscose rayon in four acid concentrations (0.01, 0.1, 1.0, and 2.5 N). The results are 25 kcal/g-mole for cotton linters, which is slightly lower than 28 kcal/g-mole obtained by Sharples [147, 150], and 32 kcal/g-mole for rayon, which is close to the values of 32, 35, and 34 kcal/g-mole for homogeneous hydrolysis of cotton linters and two kinds of wood pulp, respectively, obtained by Marchessault and Rånby [88].

Akhmetov [3] has presented a generalized mathematical model for hydrolysis of cellulose to sugar in sulfuric acid solutions; the model takes into account the initial rapid and subsequent slower hydrolyses. Unfortunately, he has not offered any test data.

Meller [96] has found that heterogeneous hydrolysis of cellulose obeyed the zero-order reaction kinetics; however, McBurney [89] has pointed out that Meller did not carry out his studies beyond a weight loss of 20% and he obtained the maximum weight loss of only 5–10 %. Thus, it is possible that at a low level of degradation, the rate would approximate a zero-order characteristic.

In short, the heterogeneous acid hydrolysis of cellulose exhibits a variety of kinetic behavior under different hydrolysis conditions employed and at different stages.

B. Diffusion Model

A process has been developed to hydrolyze cellulose catalyzed with gaseous HCl. According to Kusama [78], this process is characterized by extremely rapid hydrolysis and ready recovery of HCl; the equipment and apparatus for the process are simple.

Moiseev et al. [105] have derived a diffusion model to describe the rates of polysaccharide degradation in an acidic medium. The model is based on the data from hydrolysis of O-ethylcellulose films in vapors of aqueous hydrochloric acid of different concentrations. The model is generally written as

$$\frac{\partial C_n}{\partial t} = k\,(\overline{C}_n^0 - C_n)\,C_{HA}C_{H_2O},\qquad (4.34)$$

where C_n^0 is the initial concentration of glucosidic bonds in the polymer, C_n the concentration of hydrolyzed glucosidic bonds, and C_{HA} and C_{H_2O} the concentrations of acid and water, respectively, in the polymer. The acid concentration is calculated from the following expression;

$$\frac{\partial C_{HA}}{\partial t} = D_{HA}\,\frac{\partial^2 C_{HA}}{\partial x^2} + \sum_{i=1}^{m} C_{HA}\,C_i\,k_{pi}\qquad (4.35)$$

$$= \text{acid diffused} + \text{conjugate acid}$$
$$\quad\;\text{into the}\qquad\;\;\text{formed due to}$$
$$\quad\;\text{polymer}\qquad\;\;\text{the protonation,}$$

where D_{HA} is the effective diffusion coefficient of acid, x the diffusion coordinate, C_i the concentration of unprotonated functional group, k_p the rate constant of protonation, and m the number of protonation reactions. By means of the proton magnetic resonance (PMR) technique, Moiseev et al. [105] have determined that the polyglucosides undergo negligible protonation so that the second term in Eq. (4.35) can be ignored; thus, the equation reduces to

$$\frac{\partial C_{HA}}{\partial t} = D_{HA}\,\frac{\partial^2 C_{HA}}{\partial x^2}.\qquad (4.36)$$

The mass balance of water in a differential section of the polymer gives

$$\frac{\partial C_{H_2O}}{\partial t} = D_{H_2O}\,\frac{\partial^2 C_{H_2O}}{\partial x^2} - k\,(C_n^0 - C_n)\,C_{HA}\,C_{H_2O}\qquad (4.37)$$

$$= \text{water diffused} - \text{water consumed}$$
$$\quad\;\text{into the}\qquad\;\;\text{due to splitting}$$
$$\quad\;\text{polymer}\qquad\;\;\text{of bonds}.$$

Combination of Eqs. (4.34), (4.36), and (4.37) yields an equation for calculating the concentration of hydrolyzed glucosidic bonds in the polymer. The resulting equation, nevertheless, cannot be solved analytically without simplifications; the necessary simplifying assumptions are: (i) the diffusivities of HCl and H_2O are identical, i.e., $D = D_{HCl} = D_{H_2O}$, and D is independent of the concentration of either component in the polymer; (ii) k is constant because of the low solubility of aqueous acid in the polymer; (iii) hydrolysis of glucosidic bonds is virtually irreversible, and this condition is satisfied in a low degree of degradation (<0.02) where the concentration of hydrolyzable glucosidic bonds is almost constant, i.e., $(\overline{C}_n^0 - \overline{C}_n) \ll \overline{C}_n^0$; and (iv) the polymer properties are isotropic. O-ethylcellulose is amorphous; therefore, for the mean rate the entire volume of polymer, we have

$$C_n = k_{obs} C_{HCl} t \left\{ 1 - \frac{8}{\pi L} \sum_{m=1}^{\infty} \frac{k_{obs}(B_m^2 D + k_{obs}) + B_m^2 D \{1 - \exp[-(B_m^2 D + k_{obs})t]\}}{(2m-1) B_m (B_m^2 D + k_{obs})^2 t} \right\},$$

(4.38)

where

$k_{obs} = k\, C_n^0$,
$B_m = \pi\,(2m-1)/L$,
$L =$ diffusion distance.

The concentration of hydrolyzable glucosidic bonds, C_n, may be found from the following relationship;

$$C_n = \overline{C}_n^0 - \overline{C}_n(t) = \frac{m\,(P_n^0 - 1)}{mP_n P_n^0} - \frac{m\,(P_n - 1)}{mP_n P_n^0} = \frac{m}{mP_n} - \frac{m}{mP_n^0}$$

(4.39)

or

$$C_n = m \left(\frac{1}{M_n} - \frac{1}{M_n^0} \right),$$

(4.40)

where $\overline{C}_n^0$ and $\overline{C}_n$ are the concentrations of glucosidic bonds at the onset and time t, respectively; M_n^0 and M_n are the number-average molecular weights at the onset and time t, respectively; and m is the molecular weight of the monomer unit. Moiseev et al. [105] have found that the plot of C_n against t is linear for the degradation of O-ethylcellulose film in 28% HCl vapor at 25°, 40°, or 60 °C. In other words, the polymeric matrix is entirely saturated with the acid, and thus,

$$(B_m^2 D + k_{obs})\, t \gg 1$$

(4.41)

and Eq. (4.37) takes the following form;

$$C_n = k_{obs} C_{HCl} t.$$

(4.42)

Nomenclature

B_m $\qquad = \pi(2m-1)/L$ [Eq. (4.38)]

C_A $\qquad =$ concentration of cellulose [Eq. (4.8)]

C_B $\qquad =$ concentration of glucose [Eq. (4.9)]

C_C $\qquad =$ concentration of decomposed glucose [(Eq. 4.10)]

C_σ $\qquad =$ concentration of acid [Eqs. (4.11) and (4.12)]

c $\qquad =$ empirical constant [Eqs. (4.19) and (4.20)]

C_n $\qquad =$ concentration of hydrolyzed glucosidic bonds [Eqs. (4.34), (4.37), (4.39), (4.40), and (4.42)]

C_{HA} $\qquad =$ concentration of acid [Eqs. (4.34), (4.35), (4.36), and (4.37)]

C_{H_2O} $\qquad =$ concentration of water [Eqs. (4.34) and (4.37)]

E_1 $\qquad =$ activation energy for cellulose to glucose reaction

E_2 $\qquad =$ activation energy for glucose degradation reaction

F_L $\qquad =$ length of a folded chain [Eqs. (4.6) and (4.7)]

H $\qquad = 1 + \dfrac{C_B^0}{C_A^0}\left(\dfrac{k_{r-1}}{k_r}\right)$ [Eqs. (4.14), (4.16), (4.17), and (4.18)]

k $\qquad =$ reaction rate constant [Eqs. (4.1), (4.3), (4.4), (4.5), and (4.6)]

k_1 $\qquad =$ reaction rate constant for cellulose to glucose reaction [Eqs. (4.8), (4.9), (4.11), (4.13), and (4.15)]

k_2 $\qquad =$ reaction rate constant for glucose degradation [Eqs. (4.9), (4.10), (4.12), (4.13), (4.14), and (4.15)]

K_1 $\qquad =$ empirical constant [Eqs. (4.11) and (4.15)]

K_2 $\qquad =$ empirical constant [Eqs. (4.12) and (4.15)]

k_r $\qquad = k_1/k_2$ [Eqs. (4.14), (4.15), (4.16), (4.17), and (4.18)]

k_{obs} $\qquad = kC_n^0$ [Eq. (4.38)]

L $\qquad =$ diffusion distance [Eq. (4.38)]

m $\qquad =$ number of monomer segments in a folded chain [Eq. (4.7)]
$\qquad =$ empirical constant [Eqs. (4.11) and (4.15)]
$\qquad =$ molecular weight of the monomer unit [Eqs. (4.39) and (4.40)]

Mn $\qquad =$ number-average molecular weight at time t [Eq. (4.40)]

M_n^0 $\qquad =$ number-average molecular weight at onset of hydrolysis [Eq. (4.40)]

n $\qquad =$ total number of glucosidic bonds [Eq. (4.1)]
$\qquad =$ empirical constant [Eqs. (4.19) and (4.20)]

N $\qquad =$ number of repeating monomer units in cellulose [Eq. (4.2)]

P_n $\qquad =$ number-average degree of polymerization at time t [Eqs. (4.2), (4.3), (4.4), (4.6), (4.28), (4.29), (4.32), and (4.33)]

P_n^0 $\qquad =$ number-average degree of polymerization at onset of hydrolysis [Eqs. (4.3), (4.4), (4.6), (4.7), (4.28), (4.29), (4.31), (4.32), and (4.33)]

P_n' $\qquad =$ level-off degree of polymerization [Eq. (4.33)]

r_a $\qquad =$ rate of hydrolysis of amorphous cellulose [Eq. (4.21)]

r_c $\qquad =$ rate of hydrolysis of crystalline cellulose [Eq. (4.21)]

t $\qquad =$ hydrolysis time

W $\qquad =$ weight fraction of unhydrolyzed cellulose [Eq. (4.21)]

W_t $\qquad =$ weight of cellulose sample at time t [Eq. (4.27)]

W_0 $\qquad =$ weight of cellulose sample at onset of hydrolysis [Eq. (4.29)]

X $\qquad =$ loss-in-weight of cellulose [Eq. (4.20)]
$\qquad =$ average number of cuts per crystalline segment [Eqs. (4.31) and (4.32)]

x $\qquad =$ fractional loss-in-weight of "total" cellulose residue [Eq. (4.22)]
$\qquad =$ diffusion coordinate [Eqs. (4.35), (4.36), and (4.37)]

x_a $\qquad =$ fractional loss-in-weight in the "easy-to-access fraction" [Eq. (4.22)]

x_a' $\qquad =$ fractional loss-in-weight in the initial stage of hydrolysis [Eq. (4.24)]

x_a'' $\qquad =$ fractional loss-in-weight in the intermediate stage of hydrolysis [Eq. (4.25)]

x_a''' $\qquad =$ fractional loss-in-weight in final stage of hydrolysis [Eq. (4.26)]

Y $\qquad =$ average number of intercrystalline segments [Eqs. (4.31) and (4.32)]

144

Greek Letters

α	= number of monomer units in the largest soluble oligomer [Eqs. (4.27), (4.28), (4.29), and (4.30)]
θ_0	= rotation angle of solution at onset of hydrolysis [Eq. (4.5)]
θ	= rotation angle of solution at time t [Eq. (4.5)]
θ_∞	= rotation angle of solution at the end of hydrolysis [Eq. (4.5)]

References

1. Achwal WB et al. (1959) J Poly Sci 35:93
2. Achwal WB, Nabar GM (1960) J Text Res 30:872
3. Akhmetov KA (1975) Izo Akad Nauk Uzb SSR Ser Takn Nauk 19:78
4. Alexander WJ, Mitchell RL (1949) Anal Chem 21:1497
5. Almin KE et al. (1972) J Appl Poly Sci 16:2583
6. Andren RK et al. (1976) Appl Poly Symp 28:205
7. Atalla RH (1979) J Am Chem Soc 101:65
8. Bao Y et al. (1980) J Appl Poly Sci 25:263
9. Barth HG, Regnier FE (1980) J Chromatogr 192:275
10. Barker RH, Vail SL (1967) J Text Res 37:1077
11. Battista OA (1950) Ind Eng Chem 42:502
12. Battista OA et al. (1956) Ind Eng Chem 48:33
13. Battista OA, Coppick S (1947) J Text Res 17:419
14. Beall G, Jorgensen L (1951) ibid. 21:203
15. Beringer FM, Sands S (1953) J Am Chem Soc 75:3319
16. Betrabet SM, Paralikar KM (1978) Cell Chem Technol 12:241
17. Birtwell C et al. (1926) J Text Inst 17:1457
18. Blackwell J et al. (1977) ACS Symp Ser 48:42
19. Bose A, Tsao GT (1980) Bioconv Biochem Eng, vol 1. BERC IIT Delhi, New Delhi, p 279
20. Bose JL et al. (1971) J Appl Poly Sci 15:2999
21. Browning BL (1967) Methods of wood chemistry. Interscience, New York
22. Bunton CA et al. (1955) J Chem Soc (London) 4419
23. Caulfield DF, Steffes RA (1969) TAPPI 52:1361
24. Chang M et al. (1981) In: Fiechter A (ed) Adv Biochem Eng, vol 20. Springer, Berlin Heidelberg New York, p 15
25. Chang M (1971) J Poly Sci C 36:343
26. Chang M (1974) J Poly Sci A-1 12:1349
27. Chang M et al. (1973) J Poly Sci A-2 11:399
28. Chang M et al. (1976) National Science Council Monthly (China) 4:2665
29. Chang M (1979) Presented at the 72nd AIChE Annual Meeting, San Francisco, CA, Nov. 25–29
30. Daruwalla EH (1966) TAPPI 49:106
31. Daruwalla EH, Nabar GM (1956) J Poly Sci 20:94
32. Daruwalla EH, Shet RT (1962) J Text Res 32:942
33. Davidson GF (1943) ibid. 34:T87
34. Davidson GF, Nevell TP (1959) ibid. 50:T238
35. Davidson GF, Nevell TP ibid. 48:T356
36. Fagan RD et al. (1971) Environ Sci Technol 5:545
37. Fagerstam L et al. (1977) In: Ghose TK (ed) Proc Intern Symp Bioconv Cellulosic substances into chemicals, Energy and Microbial Protein. BERC IIT Delhi, New Delhi
38. Fan LT et al. (1980) In: Fiechter A (ed) Adv Biochem Eng, vol 14. Springer, Berlin Heidelberg New York, p 101
39. Fan LT et al. (1982) In: Fiechter A (ed) Adv Biochem Eng, vol 23. Springer, Berlin Heidelberg New York, p 158
40. Fan LT et al. (1980) In: Ghose TK (ed) Bioconv Biochem Eng, vol 1. BERC IIT Delhi, New Delhi, p 233

41. Fong WS et al. (1980) Chem Eng Prog Sept.
42. Freudenberg K (1930) Ann 460:288
43. Freudenberg K (1921) Chem Ber 54:767
44. Freudenberg K et al. (1930) ibid. 63:1610
45. Freudenberg K, Blomquist G (1936) ibid. 68:2070
46. Freudenberg K, Kuhn W (1932) ibid. 65B:484
47. Gaden EL Jr et al. (eds) (1976) Biotechnol Bioeng Symp, no 6. Interscience, New York
48. Goldstein IS (1981) Organic chemicals from biomass. CRC Press, Florida
49. Goto K et al. (1971) Agr Biol Chem 35:111
50. Grethlein HE (1978) Biotechnol Bioeng 20:503
51. Grethlein HE (1978) J Appl Chem Biotechnol 28:296
52. Halliwell G (1977) In: Ghose TK (ed) Proc Bioconv Symp. BERC IIT Delhi, New Delhi, p 81
53. Hammett LP, Paul MA (1934) J Am Chem Soc 56:830
54. Han YW (1978) Adv Appl Microbiol 23:19
55. Harris EE, Kline AJ (1949) J Phys Colloid Chem 53:344
56. Hebeish A et al. (1979) Cell Chem Technol 13:543
57. Heidt LJ, Purves CB (1944) J Am Chem Soc 66:1385
58. Hermans PH et al. (1951) Macromol Chem 6:25
59. Hermans PH, Weidinger A (1949) J Poly Sci 4:317
60. Hermans PH, Weidinger A (1951) ibid. 6:533
61. Higgin HG et al. (1958) ibid. 32:247
62. Immergut EA, Rånby BG (1956) Ind Eng Chem 48:1183
63. Isbell HS, Frush HL (1940) J Res Natl Bur Standards 24:125
64. Jellinek HHG (1944) Trans Faraday Soc 40:266
65. Jorgenson L (1950) Acta Chem Scand 4:185
66. Jorgenson L (1950) ibid. 4:658
67. Jorgenson L (1947) ibid. 3:780
68. Klages FZ (1932) Physik Chem A159:357
69. Klages FZ (1935) Ann 520:71
70. Knappert D et al. (1980) Biotechnol Bioeng 22:1449
71. Knappert D et al. (1981) Presented at "3rd Symposium on Biotechnol. in Energy Production Conservation", Gatlinburg, Tennessee
72. Kobayashi T et al. (1960) Bull Agri Chem Soc (Japan) 24:443
73. Kobayashi T (1952) Presented at the Wood Saccharification Discussion Committee, no 1, p 27
74. Kobayashi T, Sakai Y (1957) In: Asai T (ed) Koboriyokogyo. Tokyo, p 188
75. Krasag K (1976) Appl Poly Symp 28:777
76. Kuhn W (1930) Chem Ber 63:1503
77. Kusakabe IT et al. (1975) J Ferment Technol 53:135
78. Kusama J (1979) Chemical Economy & Engineering Review 11 (16):32
79. Lecttenberg VL et al. (1972) Agronomy J 64:675
80. Lee SB et al. (1983) Biotechnol Bioeng 25:33
81. Lee YH et al (1980) In: Fiechter A (ed) Adv Biochem Eng, vol 17. Springer, Berlin Heidelberg New York, p 131
82. Lin KW (1981) AIChE Symp Ser 77:102
83. Lin SY (1972) Fiber Sci Technol 5:303
84. Lokhande HT (1978) J Appl Poly Sci 22:533
85. Long FA, Paul MA (1957) Chem Revs 57:935
86. Mandels M et al. (1974) Biotechnol Bioeng 16:1471
87. Manjunath BR, Peacock N (1969) J Text Res 70
88. Marchessault RH, Rånby BG (1959) Svensk Papperstidn 62:230
89. McBurney LF (1954) In: Ott E et al. (eds) Presented at Cellulose and Cellulose Derivatives, Interscience, New York, p 99
90. McIntyre D, Long FA (1954) J Am Chem Soc 76:3240
91. McKeown JJ, Lyness WI (1960) J Poly Sci 47:9
92. McKibbin SSW (1958) Ph. D. Dissertation, University of Wisconsin
93. Meller A (1949) J Poly Sci 4:619
94. Meller A (1961) ibid. 51:100

95. Meller A (1963) ibid. C-2:97
96. Meller A (1953) ibid 10:213
97. Meller A (1951) TAPPI 34:171
98. Meller A (1952) ibid. 35:72
99. Meller A (1955) ibid. 38:682
100. Michie RIC et al. (1961) J Poly Sci 51:85
101. Millett MA et al. (1975) In: Wilke CR (ed) Biotechnol Bioeng Symp, no 5. Interscience, New York, p 193
102. Millett MA et al. (1976) In: Gaden EL Jr et al. (eds) Biotechnol Bioeng Symp, no 6. Interscience, New York, p 125
103. Millett MA et al. (1954) Ind Eng Chem 46:1493
104. Modi JR et al. (1963) J Appl Poly Sci 7:15
105. Moiseev YV et al. (1976) Carbohydrate Res 51:39
106. Montroll EW (1941) J Am Chem Soc 63:1215
107. Montroll EW (1940) J Chem Phys 8:721
108. Morehead FF (1950) J Text Res 20:549
109. Nelson ML (1960) J Poly Sci 43:351
110. Nelson ML, Conrad CM (1948) J Text Res 18:149
111. Nelson ML, Tripp VW (1953) J Poly Sci 10:557
112. Nesse N et al. (1977) Biotechnol Bioeng 19:323
113. Nevell TP, Upton WR (1976) Carbohydrate Res 49:163
114. Newth FH et al. (1947) J Chem Soc p 10
115. Nickerson RF, Habrle JA (1947) Ind Eng Chem 39:1507
116. Okamura SJ (1942) J Soc Chem Ind (Japan) 45:1104
117. Ott E, Spurlin HM (1954) Cellulose and cellulose derivatives. Interscience, New York, p 101
118. Pacsu E (1947) J Text Res 17:405
119. Parikh RS (1967) ibid. 37:538
120. Philip HL et al. (1947) ibid. 17:585
121. Planes RL (1978) Cell Chem Technol 12:355
122. Rånby BG (1961) J Poly Sci 53:131
123. Rånby BG, Marchessault RH (1956) ibid. 36:561
124. Reese ET (1976) In: Gaden EL Jr et al. (eds) Biotechnol Bioeng Symp, no 6. Interscience, New York, p 9
125. Richards GN (1955) Chem Inds (London) p 228
126. Rinaudo M et al. (1969) J Poly Sci C28:197
127. Roberts EJ et al. (1972) J Text Res 42:217
128. Roberts RS et al. (1980) In: Scott CD (ed) Biotechnol Bioeng Symp, no 10. Interscience, New York, p 125
129. Roseveare WE (1952) Ind Eng Chem 44:168
130. Roseveare WE et al. (1948) J Text Res 18:114
131. Rowland SP et al. (1971) J Poly Sci A-1 9:1623
132. Rowland SP et al. (1974) ibid. A-1 12:445
133. Rowland SP et al. (1969) J Text Res 39:530
134. Rowland SP et al. (1973) ibid. 43:351
135. Rowland SP, Roberts EJ (1972) J Poly Sci A-1 10:2447
136. Rowland SP, Roberts EJ (1972) ibid. A-1 10:867
137. Rowland SP, Roberts EJ (1974) ibid. A-1 12:2099
138. Rozmarin Gh (1977) Cell Chem Technol 11:523
139. Sabbagh NK, Fagerson IS (1976) J Chromatogr 120:55
140. Saeman JF (1945) Ind Eng Chem 37:43
141. Sakai Y (1965) Bull Chem Soc (Japan) 38:863
142. Sasaki T et al. (1979) Biotechnol Bioeng 21:1031
143. Schultz GV (1942) Z Phys Chem B52:50
144. Schultz GV, Lohmann HJ (1941) J Prakt Chem 157:238
145. Segal L (1975) Adv Chromatogr 12:31
146. Segal L, Loeb L (1960) J Poly Sci 42:341
147. Sharples A (1954) ibid. 54:913

148. Sharples A (1954) ibid. 13:393
149. Sharples A (1971) In: Bikales NM, Segal L (eds) Presented at Cellulose and Cellulose Derivatives, vol 5, pt 5. Wiley, New York, p 991
150. Sharples A (1957) Trans Faraday Soc 53:1003
151. Sharples A (1954) J Poly Sci 14:95
152. Shinouda HG, Moteleb MMA (1979) J Poly Sci Polymer Chemistry Edition 17:3329
153. Szejtli J (1975) Säurehydrolyse glycosidischer Bindungen. VEB, Leipzig
154. Thompson DR, Grethlein HE (1979) Ind Eng Chem Prod Res Dev 18:166
155. Toyama N et al. (1977) In: Ghose TK (ed) Proc Bioconv Symp. BERC IIT Delhi, New Delhi, p 373
156. Tripp VW et al. (1958) J Text Res 28:404
157. Tsao GT (1978) Proc Biochem 10:12
158. Vaux WG (1975) AIChE Meeting, Los Angeles, California, Nov. 16–20
159. Wadhera IL, Manley RStJ (1965) J Appl Poly Sci 9:2627
160. Wegner TH et al. (1982) FRRS Inds. Wood Energy Forum 82, Washington, D.C.
161. Wenzl HFJ (1970) The chemical technology of wood. Academic Press, New York
162. Whalley E (1959) Can J Chem 37:788
163. Whalley E (1959) Trans Faraday Soc 55:798
164. Wilke CR (ed) (1975) Biotechnol Bioeng Symp, no 5. Interscience, New York
165. Wolfrom ML, Snowden JC (1938) J Am Chem Soc 60:1026
166. Yorston FH (1933) Quart Rev Forest Products Laboratory of Canada 13:16

5 Design and Economic Evaluation of Cellulose Hydrolysis Processes

In this chapter, commercial hydrolysis processes for a variety of cellulosic materials are described; both enzymatically and acidically catalyzed processes are included. It appears, however, that the former remains largely at the conceptual or developmental stage; most of the commercial processes in operation, while rather limited in number, are said to belong to the latter. Nevertheless, the procedure for process economic evaluation and optimization of hydrolysis of cellulosic materials is illustrated with two examples from the former one utilizing relatively pure cellulose and the other lignocellulosic material as substrates, because of the future potential of the enzymatically catalyzed processes.

5.1 Enzymatic Hydrolysis

5.1.1 Pure Cellulosic Substrate

Economic optimization of the reactors for enzymatic hydrolysis of relatively pure cellulose is performed by Lee et al. [51]. Two types of reactors, a batch reactor and a continuously stirred tank reactor (CSTR), are considered. The economic optimization has entailed the determination of optimal hydrolysis conditions. The fractional contributions of various cost elements to the entire production cost are estimated.

In the design of both types of reactors, batch and CSTR, Lee et al. [51] have considered only the material balances. The reactors are assumed to be operating under isothermal conditions. The energy balance is neglected because the heat generated by enzyme catalyzed reactions is usually small.

The balance of any material species in a batch reactor at any moment is written as

$$\text{production} = \text{accumulation}$$

Thus for the reducing sugar

$$V_b R_P = V_b \frac{d(P)}{dt}$$

or

$$R_P = \frac{d(P)}{dt}, \tag{5.1}$$

149

where R_P is the rate of formation of reducing sugar, P and V_b is the volume of batch reactor. Similarly, for the glucose balance

$$V_b R_{P_1} = V_b \frac{d(P_1)}{dt}$$

or

$$R_{P_1} = \frac{d(P_1)}{dt}, \tag{5.2}$$

where R_{P_1} is the rate of production of glucose, P_1.

To obtain the reducing sugar and glucose concentrations for any given batch time, t_b, Eqs. (5.1) and (5.2) need be integrated simultaneously; this gives

$$(P) = \int_0^{t_b} R_P \, dt \tag{5.3}$$

$$(P_1) = \int_0^{t_b} R_{P_1} \, dt \tag{5.4}$$

If TPD denotes the desired sugar production rate in tons per day, then

$$TPD = n_b V_b (P), \tag{5.5}$$

where n_b represents the number of batches in a day. With two hours allocated to loading and unloading the reactor in each batch, the number of batches in a day is given by

$$n_b = \frac{24}{(t_b + 2)}. \tag{5.6}$$

Hence, for any given batch time t_b, the volume of the batch reactor(s) can be determined; this, in turn, gives rise to estimation of the equipment cost. For a continuously stirred tank reactor (CSTR), a steady state reducing sugar balance gives

$$V_{CSTR} = \frac{v_0 (P)}{R_p}. \tag{5.7}$$

Similarly, the steady state glucose balance yields

$$V_{CSTR} = \frac{v_0 (P_1)}{R_{P_1}}. \tag{5.8}$$

Equating Eqs. (5.7) and (5.8) yields

$$\frac{(P)}{R_P} = \frac{(P_1)}{R_{P_1}}$$

or

$$\frac{R_{P_1}}{R_P} = \frac{(P_1)}{(P)}.\tag{5.9}$$

To determine the volume of the CSTR, values of both (P) and (P_1) are needed. This can be accomplished in several steps. First, a value of (P) is assumed. In the second step, a value of (P_1) satisfying Eq. (5.9) is searched in the interval $0 \leq (P_1) \leq (P)$ for the assumed value of (P). Third, the hydrolysis rate R is calculated from the kinetic equations [50] (see Chap. 3 for details). Fourth, the hourly sugar production rate is obtained from TPD as

$$v_0(P) = \frac{TPD}{24}.\tag{5.10}$$

Finally, the volume of the CSTR, V_{CSTR} is evaluated from Eqs. (5.7) and (5.10).

The design basis of Lee et al. is summarized in Table 5.1. The sugar production rate is specified at 500 tons per day. The cellulose source is considered to be waste newspaper, which is very similar to Solka Floc. Its initial slurry concentration is assumed to vary from 10 g/l to 80 g/l. The enzyme source is considered to be the culture filtrate from *Trichoderma viride* that has a filter paper activity of 1.4 IU/l and a soluble protein (enzyme) concentration of 2.1 g/l. Furthermore, the initial enzyme concentration is assumed to vary from 0.20 to 1.0 g/l. The hydrolyzate is concentrated in multieffect evaporators, to yield a final sugar concentration of about 100 g/l, suitable for high concentration alcohol fermentation. The volume of an individual reactor is limited to 200 m³, and the suspension of the slurry is achieved by a double agitator system. The basis for estimating the product cost is summarized in Table 5.2. The equipment costs are estimated by resorting to the design equations presented in Table 5.3 [61]. When the process design dictates a larger total reactor volume than 200 m³, an integral number of equally sized reactors is used. A schematic diagram for this cellulose hydrolysis process is shown

Table 5.1. Basis of reactor design for enzymatic hydrolysis of cellulose [51]

Sugar production rate[a]	500 ton/day
Initial cellulose concentration	10–80 g/l
Initial enzyme concentration	0.25–1.0 g/l
Recovery of spent enzyme	Neglected
Recovery of undigested cellulose	Neglected
Outlet sugar was concentrated to 100 g/l by evaporation	

[a] Can supply sugar sufficient for 25 million gallon per year ethanol plant.

Table 5.2. Basis for product cost estimation [51]

Item	Cost	Remark
Raw materials		
(a) cellulose	$ 55/ton	Cellulose is free of hemicellulose and lignin
(b) enzyme	$ 2665/ton	Culture filtrate of Trichoderma reesei QM 9414
Equipment	–	10% annual depreciation in a linear fashion
Utilities		
(a) heating	$ 3/mBTU	Heat used for preheating
(b) electrical power	15¢/kwh	Electrical power for agitation and pumping
Evaporation	$ 1.4/m³ evaporated	Multiple effect evaporation for concentration of product
Labor	$ 9/h/reactor	–

Table 5.3. Cost estimation for plant equipment [61]

Item	Unit cost, $ FOB	Size Unit	Max. size
Mixing tanks (including insulation)	$482.69 (Size)^{0.514}$	Gallons	52840
Agitators	$2810 (Size)^{0.208}$	Horsepower	400
Pumps	$526.2 [2.64 + 0.0068 (Size)^{0.718}]$	PSI × GPM	–
Heat exchanger	$708 (Size)^{0.546}$	Sq. feet	–

in Fig. 5.1. The major elements contributing to reducing sugar cost include the costs of cellulose, enzyme solution, equipment, and multieffect evaporation. Thus,

Sugar cost $/g sugar
= cost of cellulose/g sugar
+ cost of enzyme solution/g sugar
+ cost of equipment/g sugar
+ cost of multieffect evaporation including both capital and operating costs/g sugar

$$= \frac{v_0 [(S)_0 - (S)] \$_c}{v_0 (P)} + \frac{v_0 (E)_0 \$_E}{v_0 (P)} + \frac{V_R}{v_0 (P)} + \left[\frac{1}{(P)} - \frac{1}{(P)^*} \right] \$_{conc}. \qquad (5.11)$$

The hydrolysis equipment include the reactors, heat exchangers, and pumps. The production cost of reducing sugar depends on the parameters $(S)_0$, $(E)_0$, and (P), as shown in Eq. (5–11). The production cost, therefore, is minimized by selecting optimal values of these parameters.

152

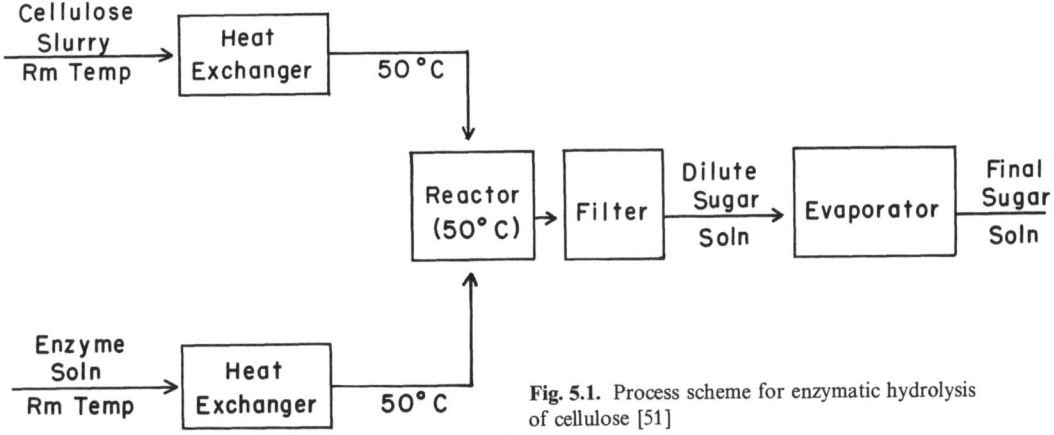

Fig. 5.1. Process scheme for enzymatic hydrolysis of cellulose [51]

Figure 5.2 depicts the production cost of reducing sugar at different enzyme and cellulose concentrations. It decreases with increases in the initial cellulose and enzyme concentrations. The lowest cost is obtained at an initial cellulose concentration of 80 g/l and an initial enzyme concentration of 1.0 g/l. The cellulose concentration of 80 g/l represents the maximum allowable cellulose concentration for homogeneous suspension of cellulose particles in the reactor. It has been known [49, 50] that beyond this level of substrate concentration, a significant fraction of the mobile phase occupies the pores in the cellulose structure. This leads to hydrodynamic instability, thus, rendering it difficult to maintain proper mixing and

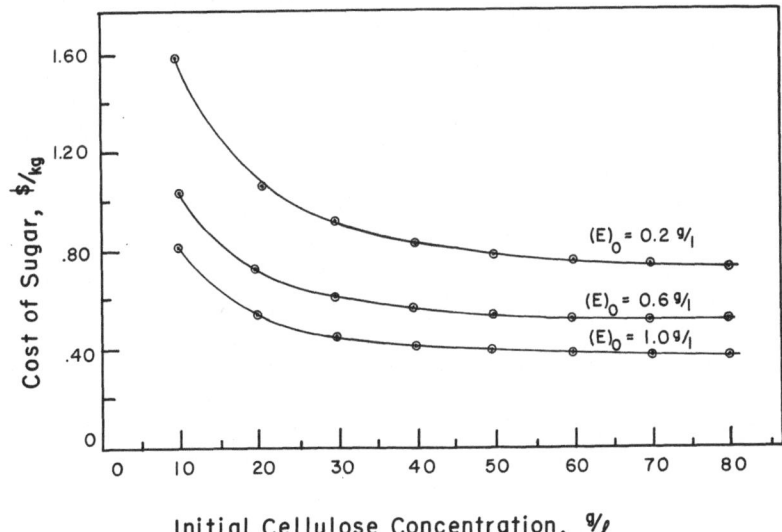

Fig. 5.2. Influence of initial enzyme and cellulose concentrations on the production cost of sugar for the batch reactor [51]

153

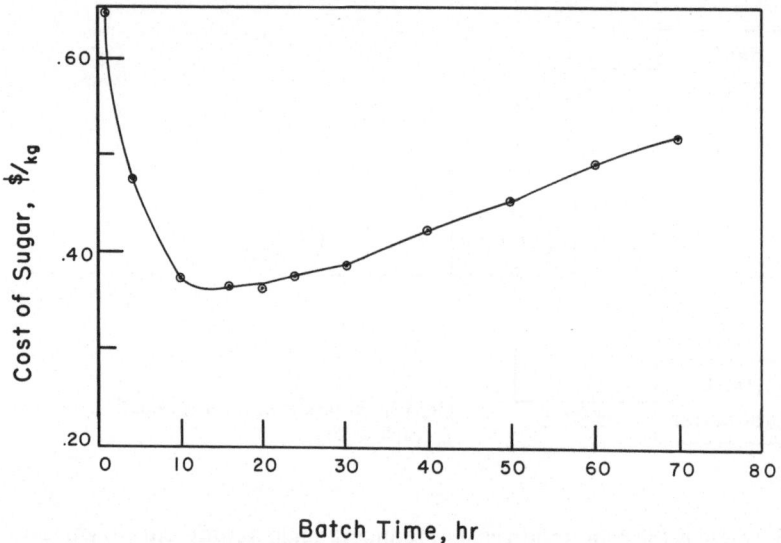

Fig. 5.3. Effect of the Batch Time on the Production Cost of Sugar; Hydrolysis Conditions, $(S)_o = 80 \, g/l$, $(E)_o = 1.0 \, g/l$ [51]

suspension of the slurry, and subsequently lowers the hydrolysis rate. The maximum initial enzyme concentration considered here is 1.0 g/l. The hydrolysis rate decreases appreciably beyond this enzyme concentration [49, 50]. This is interpreted as being due to the limited surface area for adsorption of cellulase.

The effect of the batch time on the production cost of sugar is depicted in Fig. 5.3. As can be seen in the figure the cost decreases sharply with increasing batch time up to about the first 10 h and reaches its minimum at around 16 h and then gradually increases. The hydrolysis of cellulose occurs rapidly during the first ten to fifteen hours after which it slows down gradually [48, 49]. The decrease in the hydrolysis rate can be attributed to the structural transformation of cellulose to a less digestible form, and to the inhibitory effect of the products, mainly glucose and cellobiose. The cost of sugar begins to increase steadily after about 15 h of hydrolysis because the value of incremental sugar produced after this period is exceedingly small, in comparison with the operational cost.

The optimal batch time is determined as 16 h. The corresponding final substrate concentration is 56.2 g/l, and the resultant soluble sugar concentration is 26.5 g/l of soluble reducing sugar. Although a high concentration of cellulose, namely 80 g/l, is required to minimize the product cost, it is interesting to note that only a small fraction, around 30%, is converted at the end of batch hydrolysis. The high substrate concentration increases the reaction rate and thereby reduces the reactor volume; that in turn corresponds to a reduced reactor cost. It is therefore apparent that the reduction in the reactor cost exceeds the additional expenditure incurred for providing a higher substrate concentration, a large fraction of which is wasted. It is also found that ninety batch reactors are required under the optimal batch hydrolysis conditions.

154

Table 5.4. Itemized sugar production costs for the batch reactor [51]

Cost item	¢/kg	%
Cellulose	12.52	33.18
Enzyme	5.72	15.13
Equipment	0.79	2.10
Utilities	10.95	29.03
Evaporation	3.90	10.34
Labor	3.86	10.22
Total	37.72	100.00

The contribution of each cost item to the batch production cost is summarized in Table 5.4. The raw materials contribute 48% of the total cost. The utilities' cost also contributes significantly to the overall cost.

The operation of the CSTR attains its optimum for inlet cellulose and enzyme concentrations of 32.5 g/l and 0.5 g/l respectively. The sugar concentration in the outlet stream is 23.5 g/l, corresponding to a 68.3% conversion. The optimal residence time in each reactor is 12.7 h. One hundred and twenty 200 m^3 hydrolysis reactors are required to meet the desired level of production.

The itemized production costs of sugar are summarized in Table 5.5. The cost of enzyme is the major contributor with 43.7% of the total cost. The high enzyme cost is possibly due to the inhibitory effect of the products, thereby resulting in ineffective utilization of enzyme. This inhibitory effect is more severe in the CSTR than in the batch reactor; this may result from the fact that the product concentration in the former is maintained generally at a greater level than that in the latter where it only gradually increases with time. The foregoing may be one reason for the high sugar cost of the CSTR compared to the batch reactor. A higher cost of evaporation is also observed in the CSTR than in the batch reactor. Two conflicting effects tend to counterbalance each other; according to the first the lower product concentration promotes the reaction rate by reducing the inhibition effect, thereby enhancing the utilization of enzyme, and thus achieving a lower enzyme cost;

Table 5.5. Itemized sugar production costs for the CSTR [51]

Cost item	¢/kg	%
Cellulose	6.10	9.65
Enzyme	27.44	43.70
Equipment	0.86	1.37
Utilities	14.1	22.43
Evaporation	9.85	15.69
Labor	4.50	7.16
Total	62.85	100.00

according to the second the lower product concentration from the reactor results in an increase in the cost of multieffect evaporation to satisfy the final product concentration of 100 g/l increases for lower product concentration. Again, the utilities seem to contribute significantly to the production cost; this may be attributed to the high power consumption required for maintaining a homogeneous slurry suspension.

Wilke and Mitra [80] have reported results of several conceptual design studies on enzymatic hydrolysis of newsprint. In a process employing several mixer-filter units for enzyme recovery, 885 tons/day of newsprint are passed through shredders and hammer mills to yield 200 mesh particle size (see Fig. 5.4). Out of these 885 tons, 53 tons/day of milled newsprint are sterilized with steam and diverted towards the enzyme induction. The remainder of the milled newsprint is contacted counter-currently with filtrate solution from hydrolysis for enzyme recovery in five mixer-filter stages. The hydrolysis is carried out at 50 °C and pH 4.8 for 50 h. Subsequently, the effluents are filtered in a vacuum drum filter to separate out the liquid and solid phases. The resultant net manufacturing cost of glucose is 0.01353 $/lb.

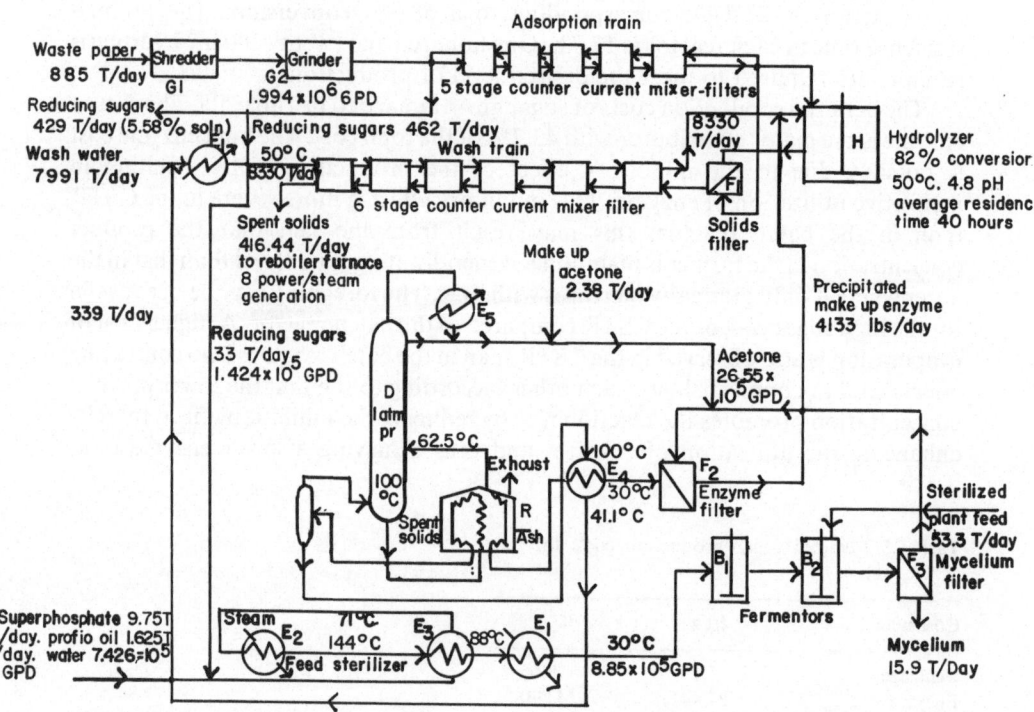

Fig. 5.4. Flow sheet of the process [80]

5.1.2 Lignocellulosic Substrate

The economic feasibility of a wheat straw hydrolysis plant employing caustic soda pretreatment is evaluated by Gharpuray et al. [26]. The pretreated wheat straw is subjected to hydrolysis in a batch reactor. The caustic soda pretreatment, despite its effectiveness in enhancing the hydrolysis rate, consumes a significant fraction of lignocellulosic biomass. For example, 40% of the cellulose, 60% of the hemicellulose, and 85% of the lignin from wheat straw are dissolved by caustic pretreatment to form black liquor (see Fig. 5.5). Special attention is paid to the processing of black liquor to utilize the dissolved biomass and also to eliminate the problems related to disposal of the alkaline black liquor by recovering the chemicals. A processing scheme that converts wheat straw to reducing sugars is proposed; it employs caustic soda for pretreatment and the resultant black liquor undergoes chemical recovery prior to its disposal. This process is optimally designed to minimize the cost of reducing sugar.

The process flow diagram is divided into two subsections. In the hydrolysis subsection, wheat straw is converted to reducing sugars, and in the caustic recovery section pretreatment chemicals are recovered (see Fig. 5.6). The milled wheat straw is first charged into the pretreatment vessels where it comes in contact with the caustic soda, forming a slurry of organic and inorganic material. This slurry is transferred to countercurrent washers, where the treated wheat straw is washed to free it from the caustic solution; eventually it is conveyed to the countercurrent enzyme recovery unit. The waste liquor emerging from the wash vessels is pumped to the caustic recovery section. The wheat straw emerging from enzyme recovery is conveyed to the hydrolysis reactors. The sugar solution from the reactors is filtered to remove unreacted wheat straw; the filtrate contains glucose as well as large part of the enzyme employed in the process. The enzyme is recovered in the

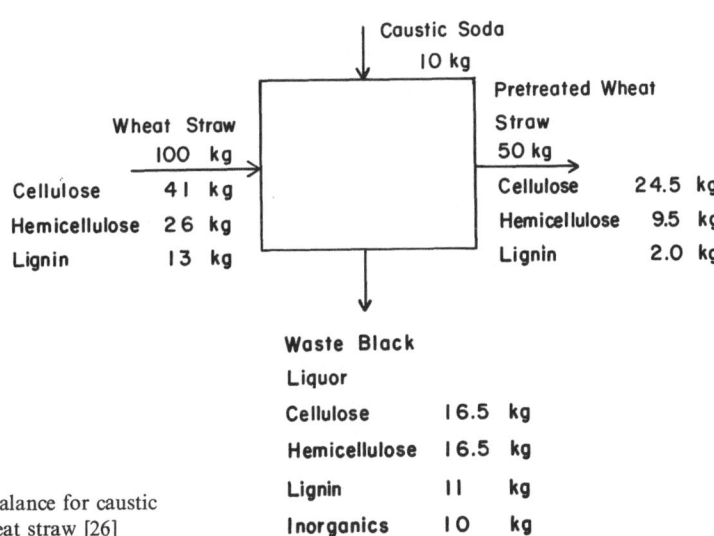

Fig. 5.5. Material balance for caustic pretreatment of wheat straw [26]

157

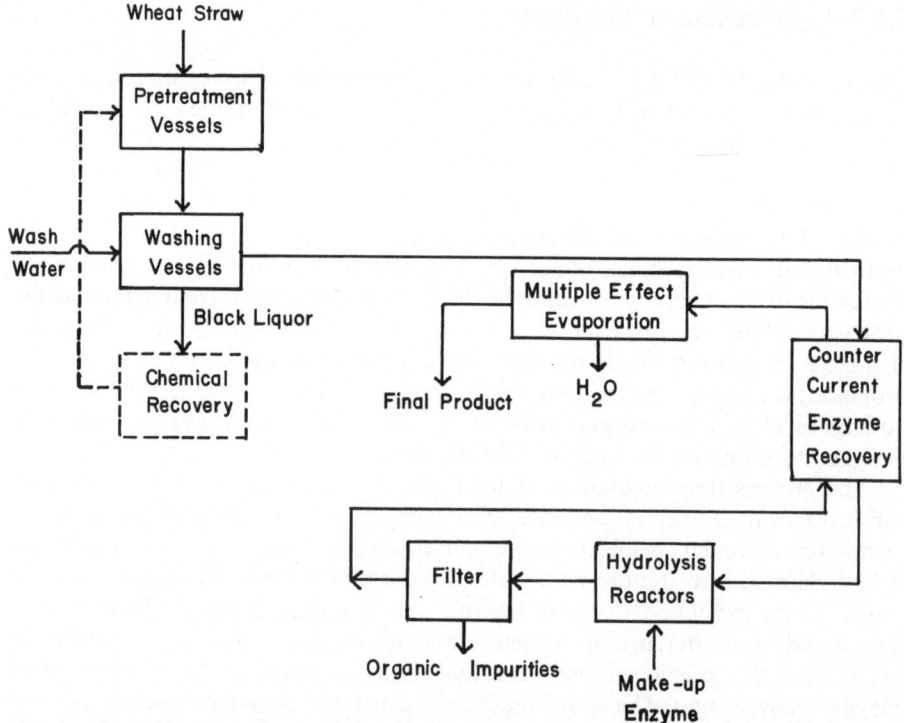

Fig. 5.6. Schematic flow diagram for a wheat straw hydrolysis plant [26]

countercurrent recovery unit where it comes in contact with pretreated wheat straw. The sugar solution is then sent to multieffect evaporators for concentration.

Emerging from the washing vessels of the overall process scheme is the black liquor containing the organic substances removed from the wheat straw and the inorganic caustic charged to the pretreatment vessels. These substances are in the form of salts and organic acids. The recovery of sodium present in these chemicals is considered essential in the overall process design. The caustic soda pretreatment of wheat straw is similar to its alkaline pulping. Therefore, a chemical recovery scheme, essentially identical to that employed in the paper and pulp industry [8] is adopted with appropriate changes (see Fig. 5.7). The black liquor containing about 15 % dissolved solids is filtered to remove the fibers. It is then conveyed to a series of seven long-tube forced circulation multieffect evaporators where the solution is concentrated to 54 % solids. The make-up sodium is added to the stream prior to its entry into the evaporators. This reduces the viscosity of the mixture in addition to minimizing the amount of scaling in the evaporators. Since the viscosity increases with increasing concentration, the flow of steam for evaporation is directed so that the hottest steam enters the last unit to contact with the most viscous fluid, while the coolest steam encounters the least viscous fluid in the first unit. The fluid exiting from the evaporator units is pumped to the recovery furnace where the organics are burnt off to leave behind a smelt containing sodium in the form of Na_2CO_3. The

158

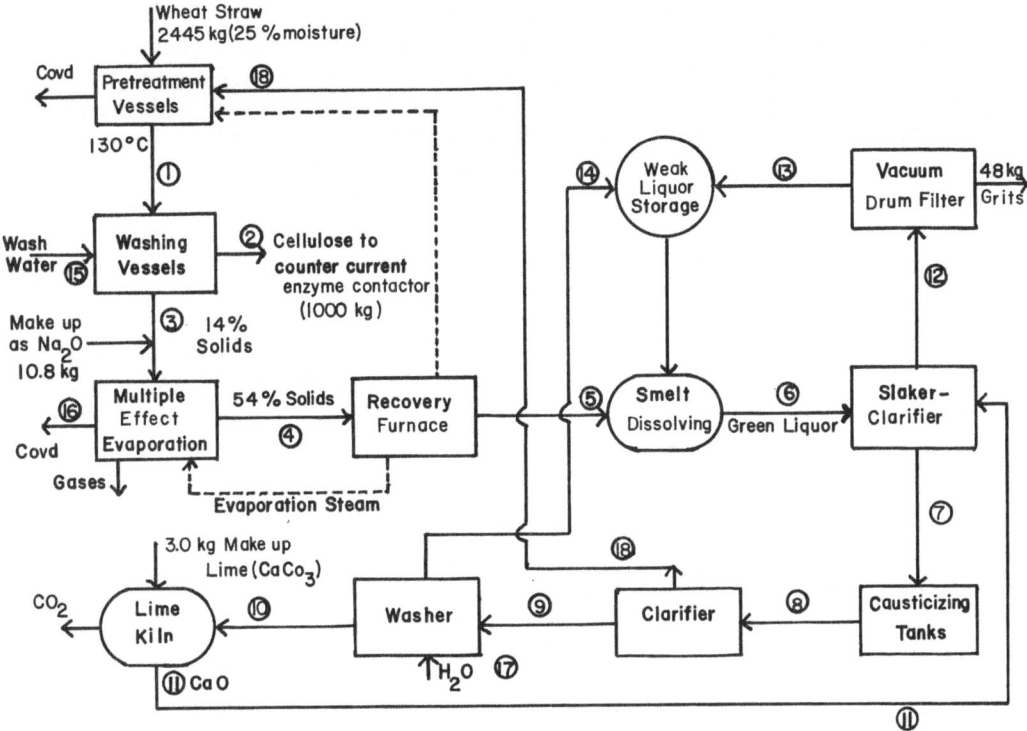

Fig. 5.7. Process flowsheet for recovery of chemicals [8]

recovery furnace is of the stationary spray type equipped with special auxiliaries for efficient black liquor solid combustion, chemical regeneration, and production of process steam. It is assumed that the heat generated in the recovery furnace is sufficient to heat both the evaporators and the pretreatment vessels.

The smelt discharged from the bottom of the furnace is dissolved in a weak liquor to form a green liquor. The green liquor is then transferred to a slaker-clarifier where it is brought in contact with lime. The rate at which lime is added is closely monitored as its reaction with water is highly exothermic and may cause shutdown if excess amounts are added. The slaker clarifier unit has the dual purpose of settling out the waste grits in solution as well as blending the lime with the aqueous mixture. The grits are sent to a vacuum drum filter and discarded, with the remaining liquid being conveyed to a weak liquor storage vessel. The lime is now in the form of $Ca(OH)_2$, and the sodium is in the form of Na_2CO_3. The mixture of these two compounds is transported to a series of three causticizing tanks with specially designed slow moving agitators to raise the efficiency of the desired reaction. The final reaction can be summarized as

$$Ca(OH)_2 + NaCO_3 \longrightarrow CaCO_3 + 2\,NaOH.$$

The resultant slurry is then pumped to a white liquor clarifier using a 35–40% underflow. From the top of the clarifier, the purified white liquor (NaOH) is

159

obtained. This is subsequently recycled back to the pretreatment vessels where it is reused for pretreatment. The bottoms exiting from the clarifier are taken to the primary and secondary lime mud washing systems where any remaining caustic is removed and pumped to the weak liquor storage vessel. The lime mud is thickened by a vacuum filter and sent to the lime kiln where make-up lime is added. The kiln is oil fired to attain a high temperature for decomposing calcium carbonate. The regenerated calcium oxide is returned to the slaker-clarifier.

The overall sodium recovery is assumed to be about 90%, and the overall lime recovery is specified to be 97.25%. Even though the recovery scheme appears to need an unnecessarily large number of processing units, it does not represent a large capital expenditure when integrated into the overall process.

The major design parameters are listed in Table 5.6. The sugar production rate is specified to be 500 tons per day. This rate is capable of supplying the substrate to a plant producing 20 million gallons of ethanol yearly. The design equations for the reactor design are listed in Table 5.7. The mathematical model of a kraft mill

Table 5.6. Basis for Reactor Design [26]

Sugar production rate[a]	500 t/day
Slurry concentration (wheat straw)	10–80 g/l
Enzyme concentration	0.2–1 g/l
Final outlet sugar concentration	100 g/l
Enzyme recovery	60%

[a] Plant supplying 20 million gallons ethanol/year.

Table 5.7. Kinetic equations for enzymatic hydrolysis of wheat straw [26, 50]

$$E_{so} = \frac{V_s \cdot (S)_o \cdot (E)_o}{(K_s + (S)_o)}$$

$$E_s = \frac{(E)_{so} \cdot (S)}{(S)_o}$$

$$\frac{DP}{DT} = K \cdot (E) \cdot (SSA)^{1.062} (100\text{-}CrI)^{0.169} (Lignin)^{-0.268} (1 - (P)/A)^2$$

$$A = 0.07 + \frac{0.67 A_A}{0.01 + A_A}$$

$$(S) = (S)_o - 0.9 (P)$$

where

$(S)_o$ = initial substrate conc., g/l
(S) = substrate conc. at any time t, g/l
$(E)_o$ = initial enzyme conc., g/l
SSA = specific surface area of substrate, m²/g
CrI = crystallinity index of substrate
Lignin = lignin content of substrate, %
(P) = product concentration at any time t, g/l
A = digestible fraction of substrate conc., g/l
$(E)_{so}$ = initial absorbed enzyme
$(E)_s$ = adsorbed enzyme at any time t
A_A = amount of caustic used for pretreatment of wheat straw, g/g wheat straw
V_s, K_s, K_o = const

Table 5.8. Basis for cost estimation [26]

Item	Cost	Remark
Wheat straw	27 $/ton	
Enzyme	2665 $/ton	
Equipment		10% annual linear depr.
Utilities		
A) heating	3 $/mBTU	
B) electrical power	15 ¢/kwh	
Evaporation	1.4 $/m^3 evaporated	Evaporation used for concentration of product
Labor	$9/reactor/h	

Table 5.9. Sugar production cost for batch reactor operation [26]

Item	¢/lb	%
Raw material costs	6.28	23.35
Equipment cost	4.29	15.95
Utilities cost	7.10	26.40
Evaporation cost	1.39	5.17
Labor cost	0.83	3.09
Pretreatment cost	7.99	29.71
Heat credit	−0.99	−3.68
Total	26.89	100.00

proposed by Boyle and Tobias [5] has been employed for modeling the chemical recovery subsection after appropriate modifications. The expressions employed for plant cost estimation are listed in Tables 5.8 and 5.9.

The values of process parameters, namely, the amount of caustic soda required for pretreatment, initial substrate concentration, initial enzyme concentration, and batch time, are determined so as to minimize the production cost of reducing sugars. The effect of substrate concentration on the production cost of sugar or simply sugar cost is represented in Fig. 5.8; the sugar cost decreases with an increase in the substrate concentration. A substrate concentration of 80 g/l is selected, which represents the maximum allowable substrate concentration so as to maintain uniform slurry suspension [49, 50]. Figure 5.9 demonstrates the influence of initial enzyme concentration on the sugar cost; again, the sugar cost decreases with an increase in the enzyme concentration. The decrease essentially ceases at an enzyme concentration of 1 g/l; therefore, this concentration is selected. It is known that beyond this concentration, the hydrolysis rate increases only slightly. The interpretation is that the surface area limits the adsorption of cellulose [49, 50]. Figure 5.10 shows the effect of caustic soda employed per unit weight of wheat straw on the sugar cost. Notice that the sugar cost drops dramatically with an increasing amount of caustic soda up to 0.05 g/g wheat straw; it increases gradually beyond this point. The plausible causes for this are as follows:

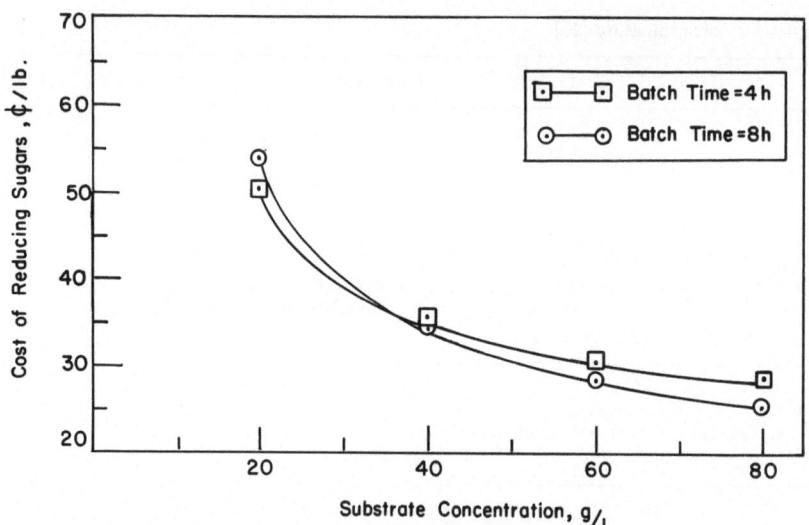

Fig. 5.8. Effect of substrate concentration on the cost of sugar [26]

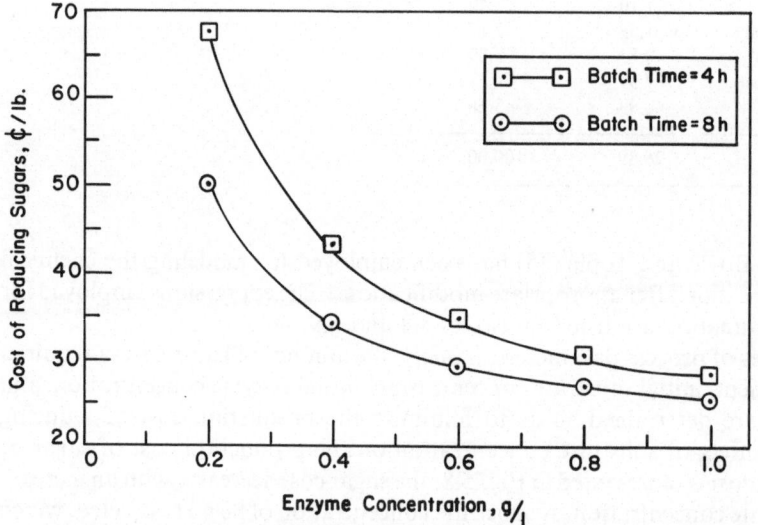

Fig. 5.9. Effect of enzyme concentration on the cost of sugar [26]

a) An increased amount of caustic soda in the pretreatment of wheat straw increases its digestibility only up to about 0.05–0.1 g caustic soda per g wheat straw; however, the incremental increase in the hydrolysis rate by application of larger quantities of caustic soda is insignificant [20].

b) Only a fraction of lignin need be removed to attain the maximum possible hydrolysis rate, as noted earlier in Chap. 3.

Figure 5.11 illustrates the influence of batch time on the cost of sugar; it decreases sharply with an increase in the hydrolysis time up to approximately 8 h

162

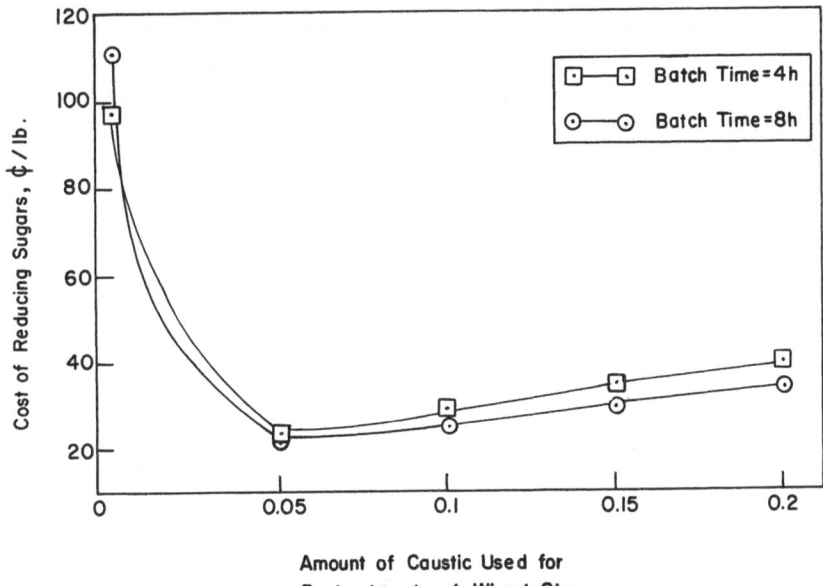

Fig. 5.10. Effect of amount of caustic soda used on the cost of sugar [26]

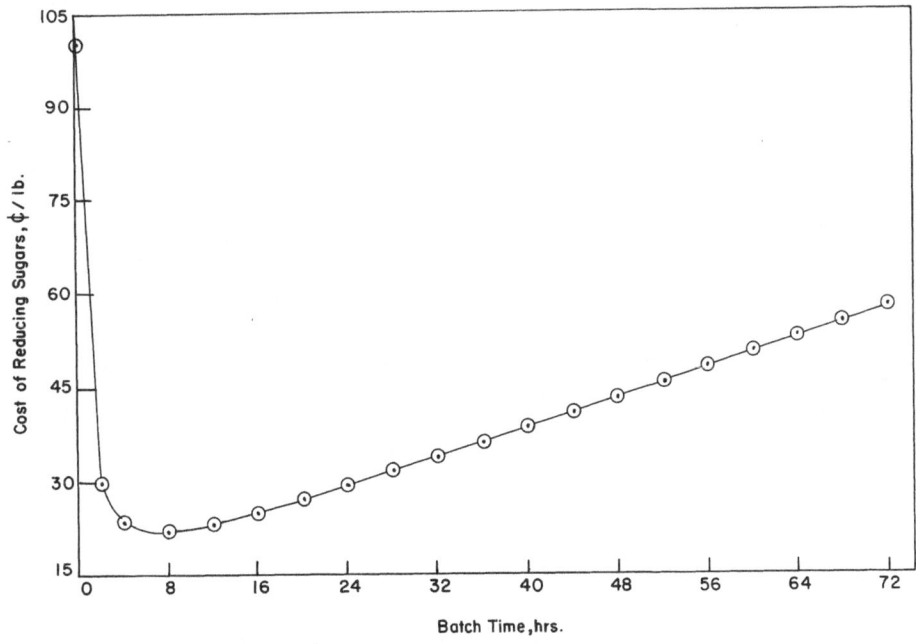

Fig. 5.11. Effect of batch time on the cost of sugar [26]

163

and then gradually increases. The hydrolysis of cellulose occurs very rapidly during the initial 6–8 h, after which it gradually decreases. This decrease may be attributed to two factors; the first is the structural transformation of cellulose to a less digestible form, and the second is the inhibitory effect of the products, mainly glucose and cellobiose, on the catalytic action of enzyme. The sugar cost starts to increase steadily after approximately 8 h because incremental sugar product beyond this period is exceedingly small, whereas the operational cost continues to accumulate. The production cost of reducing sugar is 27 ¢/lb; the contribution of each cost item towards this production cost is listed in Table 5.9. Note that raw materials represents 23% of the total production cost, utilities 26%, and pretreatment, about 30%. The high cost of utilities may be due to the heavy power required to maintain uniform slurry suspension for sustaining a high rate of hydrolysis. The relatively high cost of pretreatment is due to the caustic soda recovery, which is essential because of the copious amount of caustic soda required and because of the difficulty involved in disposing waste black liquor with high alkalinity.

5.2 Acid Hydrolysis

Basically two types or classes of acid hydrolysis processes are available for saccharification of cellulosic materials. One consists of the dilute acid-high temperature processes, and the other of the concentrated acid-low temperature processes. Sulfuric acid and hydrochloric acid have been the acids of the choice in both classes of processes.

5.2.1 Dilute Sulfuric Acid Processes

Hydrolysis of cellulosic materials with dilute acid has been carried out mainly with sulfuric acid. This class of hydrolysis processes has the advantage of ease of separating dilute sulfuric acid from the hydrolysis medium containing sugars and of the relatively low capital cost of the plant; the latter can be attributed to the fact that the cost of dilute sulfuric acid is low enough to render acid recycling unnecessary. A serious disadvantage is that the yield of glucose is usually low.

Typical saccharification processes for hydrolyzing cellulosic materials with dilute sulfuric acid are listed in Table 5.10. To minimize product contamination, the processes are usually performed in two stages. In the first, dilute acid at 130–140 °C hydrolyzes the hemicellulose portion into pentose. In the second, cellulose is hydrolyzed to glucose at an elevated temperature (170–240 °C).

Over 50 Soviet wood hydrolysis plants utilize the two-stage acid hydrolysis processes. These plants have an average annual output of 1000 tons of wood sugar per plant. Most of the product is subsequently converted into industrial alcohol and fotter yeast [36]. In the United States, a commercial plant employing a modified Madison route to process 50 ton/day of wood into alcohol has been constructed in 1982 [2].

164

Table 5.10. Processes for hydrolyzing cellulosic materials with dilute sulfuric acid

Process	Reactor	Prehydrolysis condition	Hydrolysis condition	Yield	Ref.
Scholler	Percolator	$1 \approx 2.5\%$ H_2SO_4 at 140°C	0.4% H_2SO_4 at 170°C and 8 atm	55% based on hemi- and α-cellulose	[18, 24]
Madison	Percolator	Impregnation with acid	0.5% H_2SO_4 at 185°C	72% based on hemi- and α-cellulose	[34]
T.V.A.	Percolator	0.5% H_2SO_4 at 135°C for 30 min	0.5% H_2SO_4 at 135~193°C for 50 min	72% based on hemi- and α-cellulose	[27]
N.Y.U.	Screw reactor	None	$1 \sim 1.5\%$ H_2SO_4 at 230°C for 5.25 s	50~55%	[6]
G.I.T.	Fixed bed or plug flow reactor	0.5% H_2SO_4 at 150°C	0.5% H_2SO_4 at 190°C	85%	[28, 57–59, 63]
Dartmouth	Plug flow reactor	None	1% H_2SO_4 at 235–240°C for 0.22 min	50~55%	[30, 32]
USSR	Percolator	–	–	–	[21, 44, 45, 54]
Mechanical-chemical	Percolator	1.5–5% H_2SO_4 impregnated into wood then dried	Vibratory mill, 4–80% H_2SO_4	60% based on hemi- and α-cellulose	[11–16]

165

Table 5.11. Relationship between the optimum reaction time and corresponding maximum glucose yield for the Kraft paper process at different temperatures, acid concentrations, and initial compositions [30]

Initial composition	Reaction temperature (°C)	Acid concentration (wt-% H_2SO_4)					
		0.5		1.0		2.0	
		t_{opt} (min)	C_{Bmax}	t_{opt} (min)	C_{Bmax}	t_{opt} (min)	C_{Bmax}
$C_A^0 = 1.0$	210	2.394	0.299	1.063	0.445	0.446	0.616
$C_B^0 = 0$	220	1.058	0.344	0.462	0.505	0.190	0.662
	230	0.481	0.390	0.206	0.553	0.0834	0.703
	240	0.224	0.435	0.094	0.598	0.0377	0.740
	250	0.107	0.480	0.0444	0.639	0.0175	0.772
	300	0.00373	0.670	0.00144	0.796	0.00054	0.883
$C_A^0 = 0.9$	210	2.059	0.302	0.970	0.456	0.419	0.617
$C_B^0 = 0.1$	220	0.932	0.346	0.426	0.506	0.180	0.663
	230	0.431	0.392	0.192	0.554	0.0793	0.704
	240	0.204	0.437	0.0886	0.598	0.0360	0.740
	250	0.0983	0.481	0.0419	0.640	0.0167	0.773
	300	0.00353	0.671	0.00139	0.796	0.00052	0.883
$C_A^0 = 0.8$	210	1.615	0.312	0.859	0.462	0.388	0.620
$C_B^0 = 0.2$	220	0.771	0.355	0.384	0.511	0.168	0.665
	230	0.369	0.399	0.175	0.588	0.0747	0.706
	240	0.179	0.443	0.0818	0.602	0.0340	0.742
	250	0.0879	0.487	0.0390	0.643	0.0159	0.774
	300	0.00331	0.674	0.00133	0.797	0.000503	0.884

According to kinetics, increases in temperature and pressure favorably enhance the rate of hydrolysis and yield of glucose [23]. For example, in the economical range of acid concentration, namely, 0.5–2.0%, hydrolysis of cellulose proceeds at a sufficiently high rate only when the temperature exceeds 180 °C [66]. In general, the higher the temperature, the higher the yield, as shown in Table 5.11 [30]; however, an optimum set of operating conditions exists. From Fig. 5.12 and Table 5.11, it is apparent that a short reaction time favors the yield of sugars, but exceedingly short operating time is difficult to implement. On the other hand, the yield of glucose increases with an increase in the acid concentration, the optimum acid concentration is determined on the basis of economic considerations. In general, the reaction time of less than 0.1 min and the acid concentration over 2% are impractical [30]. An ideal process operates continuously provides an isothermal reaction environment, and maintains an optimum residence time under the optimum temperature [30]. A laboratory-scale continuous plug flow reactor (PFR) was employed for acid hydrolysis of Solka Floc [31]. Hydrolysis was carried out in a 0.4–1.2% sulfuric acid solution at a residence time of 0.22 min and a temperature of 235 °C; the conversion was as high as 55% of the potential glucose [31]. A pilot-plant design was completed by using the kinetic model with Fagan's parameters [19]. Larger pilot plant scale continuous processes have been applied for acid hydrolysis of cellulosic materials [32, 33, 76, 77]. Several types of reactors such as twin screw reactors [6], and fixed and plug flow reactors [30, 57–59, 63], have been used to enhance the yield of sugars. According to Song and Lee [73], the counter-current reactor is preferable to reactors of other types, such as conventional percolation, co-current, and plug-flow reactors. Rowland et al. [65] have designed a

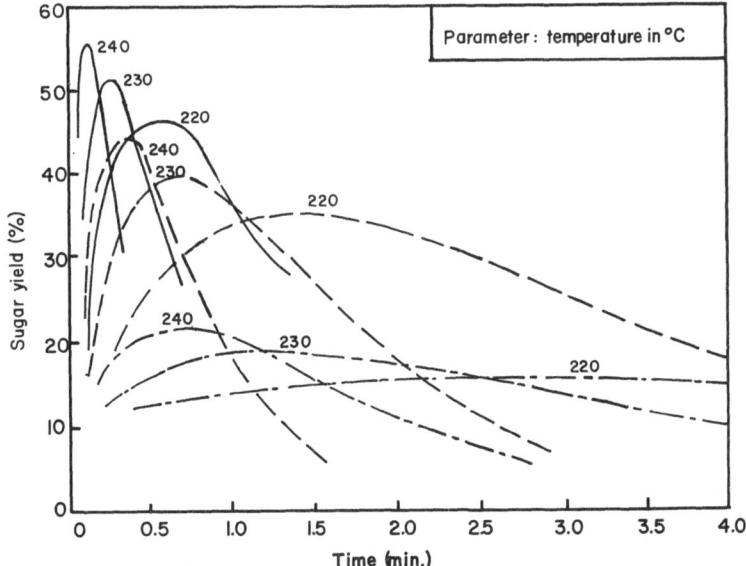

Fig. 5.12. Dependence of the predicted sugar yield on potential glucose and hydrolysis time; —, 1% acid; – – –, 0.5% acid; –·–, 0.2% acid [23, 30]

specific apparatus which promotes saturation of accessible surfaces of cellulose particles and removal of inhibiting byproducts from the reacting environment. In a method developed by Sakai [67], hydrolysis has been carried out in solutions of sulfur trioxide having their equivalent sulfur dioxide contents in the range of 0.32 to 3.31%. It has been pointed out that this method may be applied to the saccharification of wood.

The capital cost for a percolation hydrolysis plant has been estimated by Wright and d'Agincourt [82]. The sugar production capacity of this plant is such that it can supply enough sugar to produce yearly 50 million gallons of ethanol. The cost estimation is presented in Table 5.12. It can be observed that the hydrolyzers (hydrolysis reactors) are the dominant cost item, accounting for 75% of the total cost. This cost is high because of the acid brick lining required for acid resistance. A similar estimate for a plug flow reactor is given in Table 5.13. The most important cost items are the solids pumps and the reactors, which together account for 61% of the cost. The centrifuges account for 14% of the cost. Another major expense is for the various screw conveyors necessary for conveying the solids (12% of the total).

An optimization study was performed by Roberts et al. [64] to design a cellulose-recycle reactor system for producing hexose at a minimum cost. They assumed that the process feed is essentially pure cellulose and the kinetic model proposed by Saeman [66] is applicable.

Clements et al. [17] have estimated the cost of ethanol produced from cotton gin residue upon dilute acid hydrolysis and fermentation. Their analysis is presented in Table 5.14. Note that the production capacity has substantial influence on the cost of ethanol.

Table 5.12. Capital cost estimate for percolation hydrolysis [82]

Description	Purchased equipment cost (K$)
Percolation hydrolysis reactors (20)	4,950,000
Wood chip storage bin	74,000
Conveyor	110,000
Acid metering pump	15,000
Lignin receiver	92,000
Flash tank	62,000
Heat exchangers (2)	768,000
Flash tank	48,000
Heat exchanger	264,000
Feed water pump	73,000
Solubles holding tank	317,000
Total	6,773,000

For production of 50 million gal ethanol/yr.

Table 5.13. Capital cost summary for plug-flow hydrolysis [82]

Description	Purchased equipment cost (K $)
Plug-flow reactor (7)	370,000
Feed screw conveyor	129,500
Feed mixer	297,200
Reactor feed pump (7)	3,150,000
Reactor flash tank	65,000
Screw conveyor	102,600
Centrifuge	414,600
Screw conveyor	79,800
Repulping mixer	156,300
Screw conveyor	91,300
Centrifuge	414,600
Screw conveyor	36,000
Screw conveyor	238,000
Overflow pumps (2)	31,500
Polishing filter	109,500
Screw conveyor	20,000
Wash water tank	13,300
Wash water pump (2)	8,700
Filtrate tank	43,300
Filtrate pump	19,400
Total	5,790,600

For production of 50 million gal ethanol/yr.

Table 5.14. Economies of size for cotton gin residue alcohol [17]

Item	500 Gal.	1,000 Gal.	3,000 Gal.
Interest	$ 0.364	$ 22.500	$ 13.977
Maintenance	8.97	6.000	3.727
Labor	82.735	47.389	24.291
Utilities	13.485	13.485	13.485
Acid	11.818	11.818	11.818
Denaturing agent	1.500	1.500	1.500
Yeast	0.970	0.970	0.970
Insurance and bonding	2.230	1.655	1.030
Transportation and handling	11.050	9.800	9.640
Feedstock	39.600	39.600	39.600
Miscellaneous	6.061	4.545	2.525
Total ($/100 gallon)[a]	$207.91	$159.26	$122.55

[a] Excluding accelerated capital recovery costs.

5.2.2 Concentrated Sulfuric Acid Processes

Crystalline cellulose and natural hemicellulose can be dissolved completely by sulfuric acid having concentration larger than 70 % at room temperature. In such a process, hydrolysis is carried out essentially in a homogeneous medium; thus, the process is often employed in the laboratory as a quantitative saccharification technique for determining potential glucose in cellulose. Commercial implementation of the process may be difficult, however, because of the high cost of concentrated sulfuric acid and its subsequent recovery. According to information available in the open literature, only one such plant has been built in Japan, and it employed the "Hokkaido process" [40, 60]. In this process acid recovery and use of $CaSO_4$ were coupled. Other similar processes reported in the literature are listed in Table 5.15. These processes consist of the following three stages: (1) prehydrolysis to hydrolyze the hemicellulose portion, (2) main hydrolysis to hydrolyze the α-cellulose, and (3) posthydrolysis to hydrolyze oligosaccharides formed in step (2). One such process is described below [1, 53].

The chopped corn stover feed is mixed with a recycled solution of sulfuric acid and sugars (primarily glucose). The wet solids are conveyed to the prehydrolysis tank where they are heated to 100 °C (212 °F) with steam and held for two hours while the hemicellulose is hydrolyzed to glucose, xylose, arabinose, and acetic acid. The relatively mild reaction conditions lead to hemicellulose breakdown but do not degrade the resultant sugars or crystalline cellulose. Therefore, almost stoichiometric yields of sugars are obtained from the hemicellulose and amorphous cellulose.

After the prehydrolysis, the resultant sugars are removed by leaching. The solids are washed three times in the hydrolysis vessel. The first wash produces the most concentrated sugar solution and is the product stream. The effluent of the second and third washes are dilute sugar streams and are recycled. The second wash effluent is used for the first wash of the next batch, and the effluent of the third wash is used to provide the second wash. In this manner, an operation carried out in a batch manner gives the same results as continuous countercurrent leaching.

After the third wash, lignocellulosic solids (LIC) are removed from the hydrolysis/leaching vessel and centrifuged to remove much of the water. The solids leave the centrifuge operation with a moisture content of 55 %. The solids are mixed with water and acid to form a slurry and pumped to a soak tank where the slurry is soaked for two hours to provide time for the acid to permeate uniformly into the fibers. A low temperature is maintained to prevent significant hydrolysis from occurring. The slurry is centrifuged and the liquid stream recycled. The separated lignocellulosic solids, at a moisture content of approximately 55 %, are transported to the eight rotary dryers. Here the solids are dried to a moisture content of approximately 10 % at 85 °C. This operation requires roughly 1100 Btu per pound of solids processed. The drying removes water but not the sulfuric acid. This means that the cellulose becomes permeated with concentrated sulfuric acid (the ratio of acid to acid plus water becomes 0.83). Concentrated acid disrupts the lattice of crystalline cellulose by breaking the hydrogen bonds between adjacent cellulose chains. The dry acid-impregnated solids are then mixed with water and heated to 140 °C with steam. The cellulose is hydrolyzed subsequently to glucose in a few

Table **5.15.** Hydrolysis of cellulose with concentrated sulfuric acid

Process	Reactor	Prehydrolysis	Mainhydrolysis	Posthydrolysis	Yield	Ref.
Hokkaido	Flash mixer	Steam at 185 °C for 2 h to form furfural or 1.2–1.5 % H_2SO_4 at 150 °C to form xylose	80 % H_2SO_4 at room temperature	5–10 % H_2SO_4 at 100 °C, 100–220 min	40 % based on hemicellulose 90 % on α-cellulose	[40, 60]
Arkansas	Plug flow reactor	4.4 % H_2SO_4 at 100 °C, 60 min	Impregnation with 85 % H_2SO_4 at 110 °C, 10 min	8 % H_2SO_4 at 100 °C, 10 min	95 % on hemi- and α-cellulose	[1, 53]
LORRE	BSTR and CSTR	Dilute H_2SO_4	70 % H_2SO_4	dilute H_2SO_4	–	[25, 47]
Giordani-Leone	Edgerunner	0.5–1 % H_2SO_4 at 15 °C, 50 min	78 % H_2SO_4 at 40 °C, 15 min	20 % H_2SO_4 with steam, 30 min	–	[9]
Drying sacch-arification	Through dryer	45 % H_2SO_4	Drying at 30 °C	8.4 % H_2SO_4 with steam	67 % sugar	[43]

171

minutes at a low temperature and with minimal formation of by-products. The processing steps following the drying are relatively fast and are performed in a continuous manner.

The product material from the hydrolysis reactor is discharged to a flash drum. The solids and liquids are separated in a centrifuge, and the solid cake is washed to remove residual sugars. The liquid stream comprising water, sugars, and acid, is recycled to prehydrolysis, while the solids are transported to waste heat generator where they are burned to provide the energy for the process.

The sugars produced in both the hydrolysis and prehydrolysis reactors are removed in the leaching process. The resultant product stream has a composition of 7.0 wt-% glucose and 5.5 wt-% xylose. The ratio of sugars produced to acid consumed is 1.8 based on glucose, and 3.5 based on glucose, xylose, and arabinose. The ratio of water used to sugars (glucose, xylose, and arabinose) produced is 8.2.

The Laboratory of Renewable Resources Engineering, Purdue University (LORRE) has also developed a similar process for converting corn stover to sugars. The hydrolysis plant has been sized to produce 454 tons per day of C_6 sugar from 1417 tons per day of corn stover (dry basis). Prehydrolysis with dilute acid is carried out in batch stirred-tank reactors; three continuous stirred-tank reactors are used for the main hydrolysis with 70% sulfuric acid. This approach appears to possess some desirable features. For example, a stream containing C_5 sugars from prehydrolysis and a stream containing C_6 sugars from main hydrolysis may be separately produced; decomposition of sugars produced could be minimized so as to increase the yields and efficiency in the subsequent fermentation of these sugars to alcohols and other organic products [25, 47].

Hydrolysis of wood with concentrated sulfuric acid is not only an ancient practice but also a subject of current interest. This is because the concentrated acid, low-temperature processes may yield sugars almost quantitatively and may contribute towards development of biochemical processes. In order to commercialize this hydrolysis process, the manufacturing steps must be simplified, the energy consumption reduced, and difficulties encountered in recycling spent acids eliminated. A variety of new reactors have been proposed for main hydrolysis [9, 41, 42, 72, 75]. The dialysis with anion-exchange membranes has been utilized for producing fertilizers [38, 39, 56]. A saccharification method has been proposed recently for wood chips impregnated with 46% sulfuric acid by a drying operation using the vapor from petroleum fractionation [29].

5.2.3 Hydrochloric Acid Processes

Hydrolysis of cellulosic materials with concentrated acid has been carried out mainly with hydrochloric acid. The reason is that hydrochloric acid is volatile and therefore, it can be recovered easily by vacuum evaporation. Typical concentrated hydrochloric acid processes for the saccharification of wood cellulose are listed in Table 5.16. Bergius was apparently the first to develop a large-scale saccharification process employing concentrated hydrochloric acid. The acid was recovered by vacuum stripping [79].

Table 5.16. Hydrolysis of cellulose with hydrochloric acid.

Process	Reactor type	Prehydrolysis condition	Mainhydrolysis condition	Posthydrolysis condition	Yield	Ref.
Rheinau	–	Impregnated with water or dilute HCl	Cold fuming HCl	None	–	[81]
Rheinau-Bergius	Diffusion battery	None	41% HCl 3:1 acid-wood ratio	None	70% based on hemi- and α-cellulose	[79]
Modified Rheinau	A series of towers	1% HCl to hydrolyze hemicellulose to pentoses at 130°C	41% HCl	Dilute acid	85% based theoretical value	[68]
Udio-Rheinau	A series of towers	32% HCl at 20°C	41% HCl	12% HCl	29% crystalline sugars based on dry wood	[62]
Prodor	Stirred tanks with 12 shelves	–	Countercurrent contact with HCl gas over 8 h	–	–	[78]
Hereng	Vessel with inclined trays	Moist wood chips impregnated with 30% HCl, 45 min	Recontact with 30% HCl followed by countercurrent HCl gas contacting	Conditions not given	–	[21]
Noguchi-Chisso	Fluidized bed	5% HCl, 100°C, 3 hr 0.75 acid and wood ratio	Stagewise, countercurrent contact with HCl gas, at 5–125°C	1% HCl, 122°C 30–45 min	23% on hemi-cellulose and 95% on α-cellulose	[52]
Chalov	Fluidized bed	Impregnated with 43.5 HCl 0.9~1.1 wood and acid ratio	Saturated with HCl gas and then hydrolyzed at 60°C for 6.5h	–	Nearly theoretical yield of sugars	[10, 16]
Mechanical-chemical	Semi-continuous vertical tanks	Grinding equipment	Grinding in the presence of 5% HCl for 40 min	0.2% HCl at 150°C 80 min	90% on hemi- and α-cellulose	[69]

After World War II, an improved concentrated hydrochloric acid process was designed. This is the so-called "Riehm process" [62] wherein prehydrolysis is carried out in a tower-type reactor. The prehydrolyzate is transferred to another tower-type reactor wherein main hydrolysis is carried out with 41 % hydrochloric acid. The hydrochloric acid is recycled through a three stage recovery system, and its concentration is adjusted for both prehydrolysis and main hydrolysis. Finally, the sugar solution is diluted down to 12 % and subjected to posthydrolysis.

Disadvantages of this process are severe requirements for material of construction and for relatively large reactor volume per unit production [30]. The possibility of high glucose yield has been the impetus despite these disadvantages.

No sharp distinction can be drawn between the processes employing concentrated hydrochloric acid and those with hydrogen chloride gas [79]. Hydrogen chloride gas is used not only to shift the hydrolysis into the interior of the wood particles but also to facilitate the recovery of the acid. Based on this principle, several methods have been proposed. They include the Prodor process [78], the Hereng process [21], the Noguchi-Chisso process [52], and the Chalov process [10, 16]. Among them, the Chalov and Noguchi-Chisso processes are very similar: the wood is impregnated with concentrated hydrochloric acid before being treated with hydrogen chloride gas. A process developed by Sharkov et al. [70] introduces the hydrogen chloride gas at 10–20 atm in an autoclave containing dry cellulosic material (5 % moisture). A plant for implementing such a process has been built in the USSR [22]. Bose et al. [4] have demonstrated that the treatment with gaseous hydrogen chloride at 25 °C under ordinary pressure, prior to hydrolysis with concentrated hydrochloric acid, enhances the saccharification yield of native cellulose and groundnut shell pulp.

In the so-called flash saccharification process of Kusama [46], particles of wood in a semi-dry state impregnated with concentrated hydrochloric acid are suspended in a high speed hydrogen chloride gas to complete the hydrolysis reaction. The wood particles are heated thereafter to a higher temperature to recover hydrochloric acid and hydrogen chloride gas.

5.2.4 Hydrofluoric Acid Process

A process for saccharification of cellulose using hydrofluoric acid has been proposed by researchers at Michigan State University [3]. Their contention is that the process offers advantages such as minimal feedstock pretreatment, high glucose yield, low chemical costs, and undamaged lignin.

In the process, small chips of aspen wood are dried and then treated with anhydrous hydrogen fluoride for approximately an hour in vacuum distillation apparatus. This reaction yields glucosyl fluoride which subsequently is converted into glucose and hydrogen fluoride via reaction with water. The hydrogen fluoride is recovered either by evacuation or by applying low temperature heat. Water is then added to the residue to form a water-soluble sugar fraction and an insoluble lignin fraction; the latter is separated by filtration or centrifugation. Sugar yields are said to range from 45 to 99 % of theoretical. However, a large fraction of the sugar monomers formed during the solvolysis stage recombine to form oligomers during the process of hydrogen fluoride removal. Fortunately, more than 90% of these oligomers can be reconverted to the monomer form by posthydrolysis treatment for one hour at 140°C with either dilute sulfuric acid or dilute hydrofluoric acid.

A mechanical-chemical method has been proposed by Sharkov et al. [69]. The optimum conditions for this process require a short grinding period with high power input and prevention of prolonged exposure to a high temperature. Subsequently, Tanaka et al. [74] have studied a combined system of acid hydrolysis and enzymatic degradation. The former is characterized by its high rate and the latter by its high substrate specificity.

It is known that hydrolysis with dilute acid is suitable only for fermentation, whereas concentrated acid processes, resulting in relatively high yields of sugars, are suitable for production of the crystalline form of glucose. The acid from this process, however, is difficult to recover. A process with concentrated hydrochloric acid has been industrialized in Germany.

A process with hydrogen chloride gas is a special case of the process with concentrated hydrochloric acid; nevertheless, it is simpler than that with hydrochloric acid in operation and plant design, and is suitable for treating sawdust [46].

Flow sheets of typical processes for acid-catalyzed hydrolysis of cellulose are summarized in Fig. 5.13 through 5.25. The general trend of process development appears to emphasize the reduction in residence or contact time in the hydrolysis reactor; the short contact time tends to minimize the reactor size and formation of by-products which hinder subsequent fermentation of the hydrolyzate.

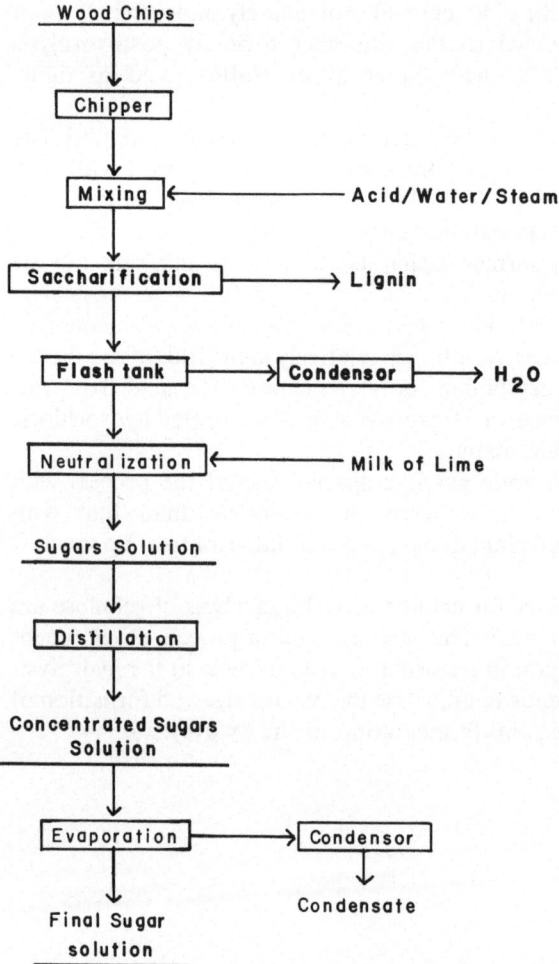

Fig. 5.13. Madison wood saccharification process [79]

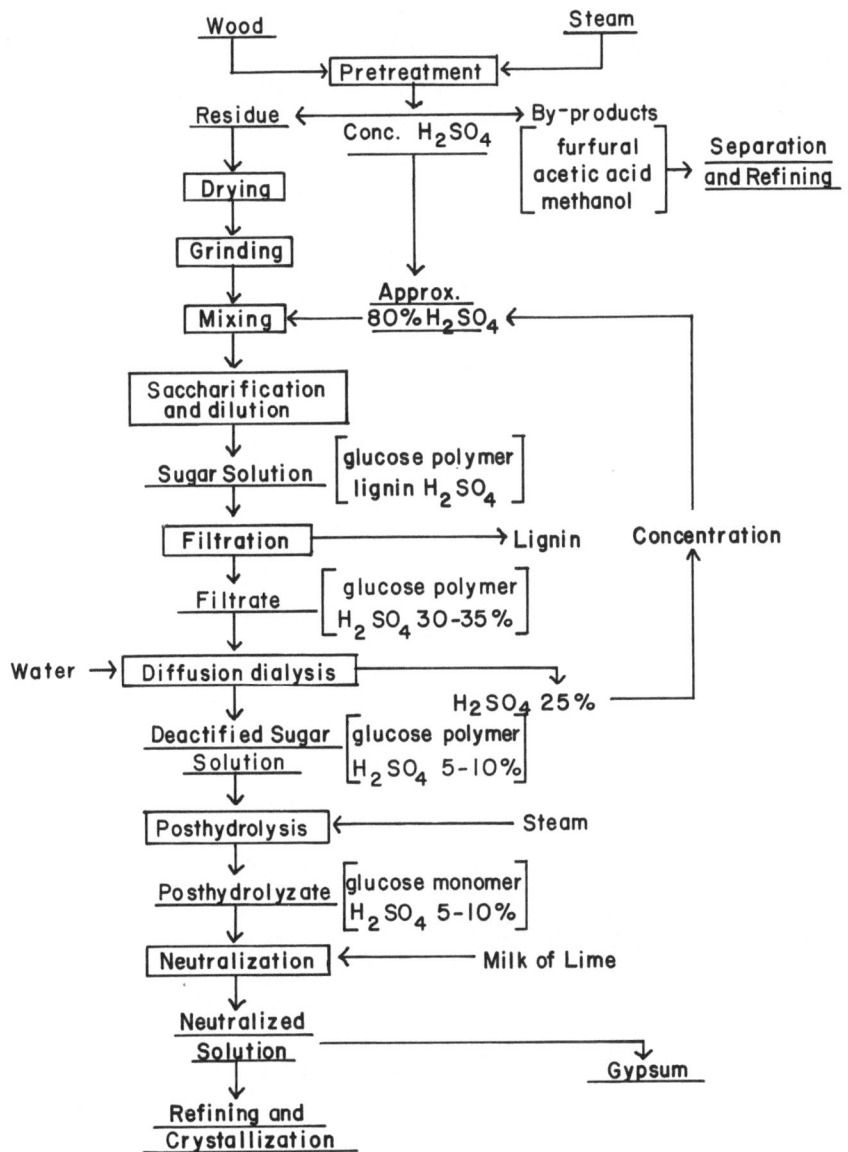

Fig. 5.14. Hokkaido wood saccharification process [79]

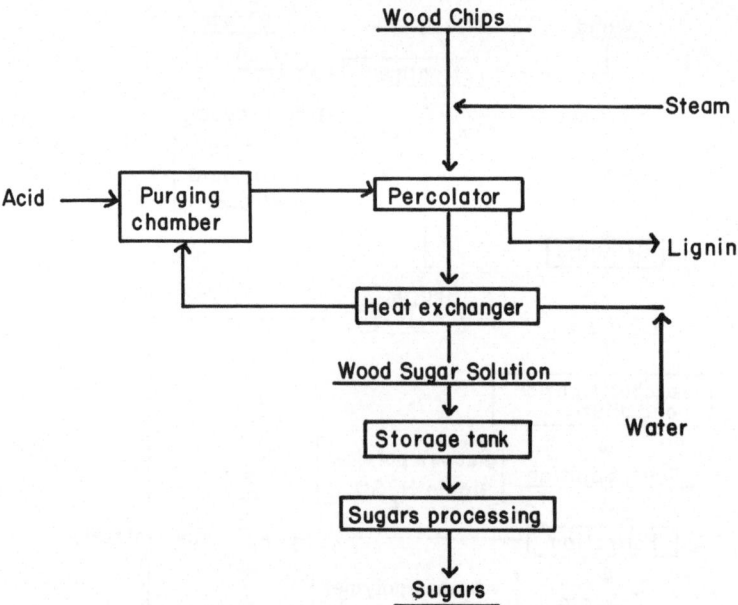

Fig. 5.15. Scholler-Tornesch wood saccharification process [79]

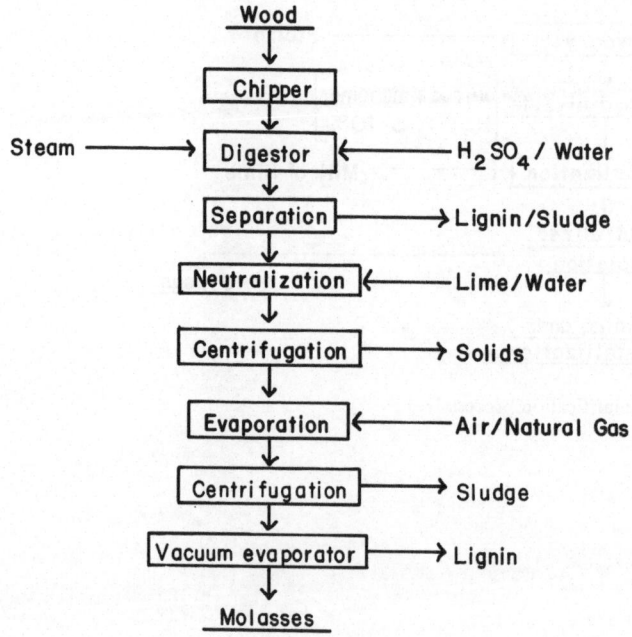

Fig. 5.16. TVA wood saccharification process [6]

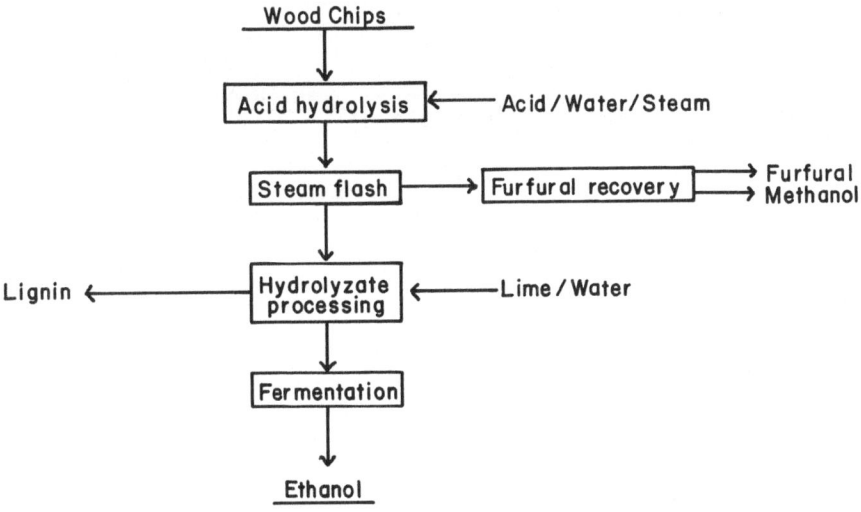

Fig. 5.17. NYU/EPA acid hydrolysis process to produce sugars for fermentation [37]

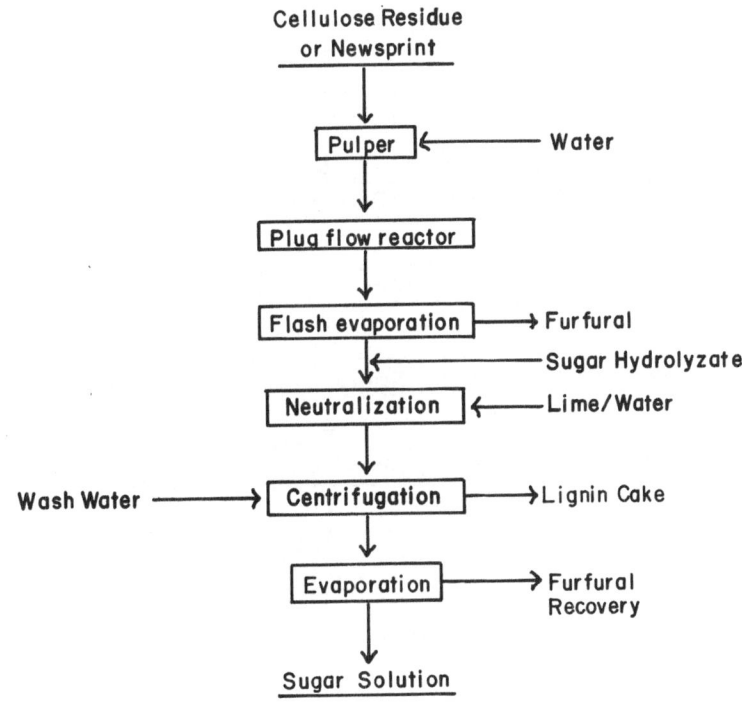

Fig. 5.18. Dartmouth University newsprint saccharification process [76, 77]

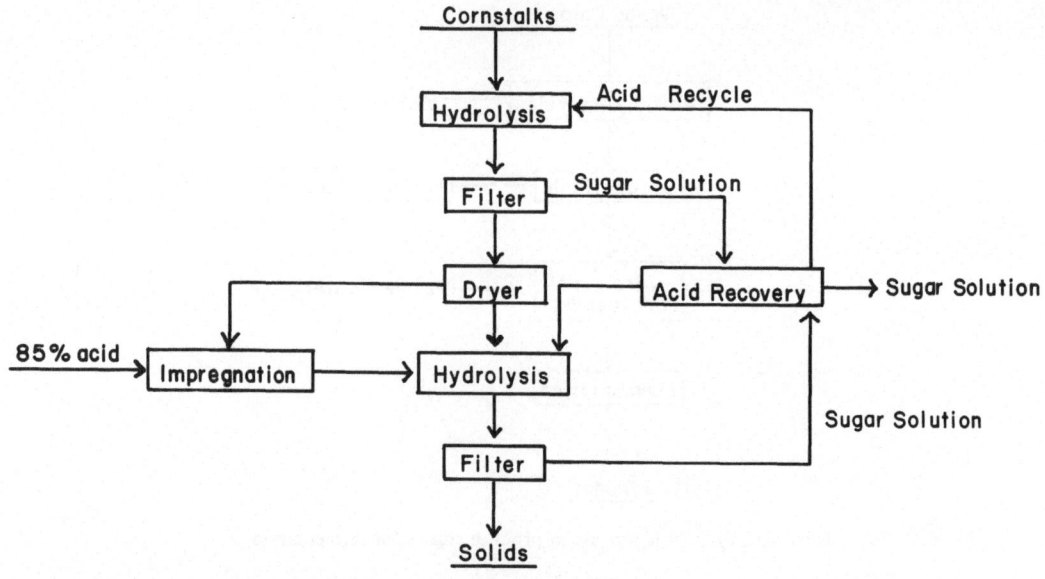

Fig. 5.19. University of Arkansas corn residue saccharification process [71]

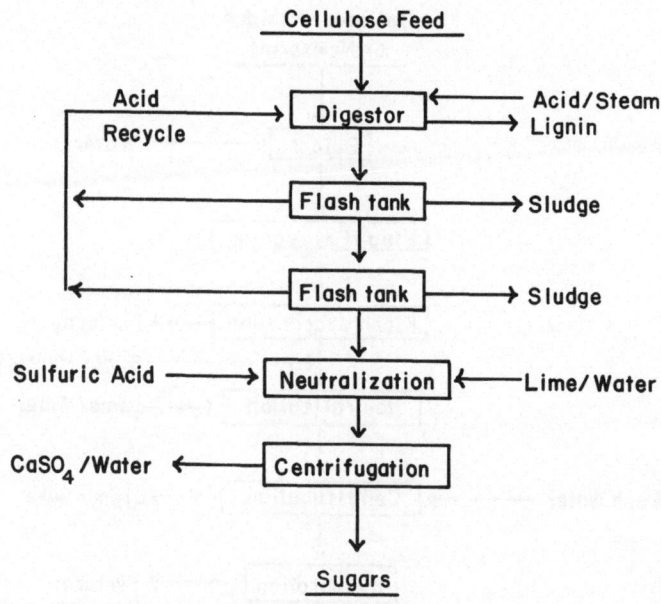

Fig. 5.20. IONICS batch saccharification process [55]

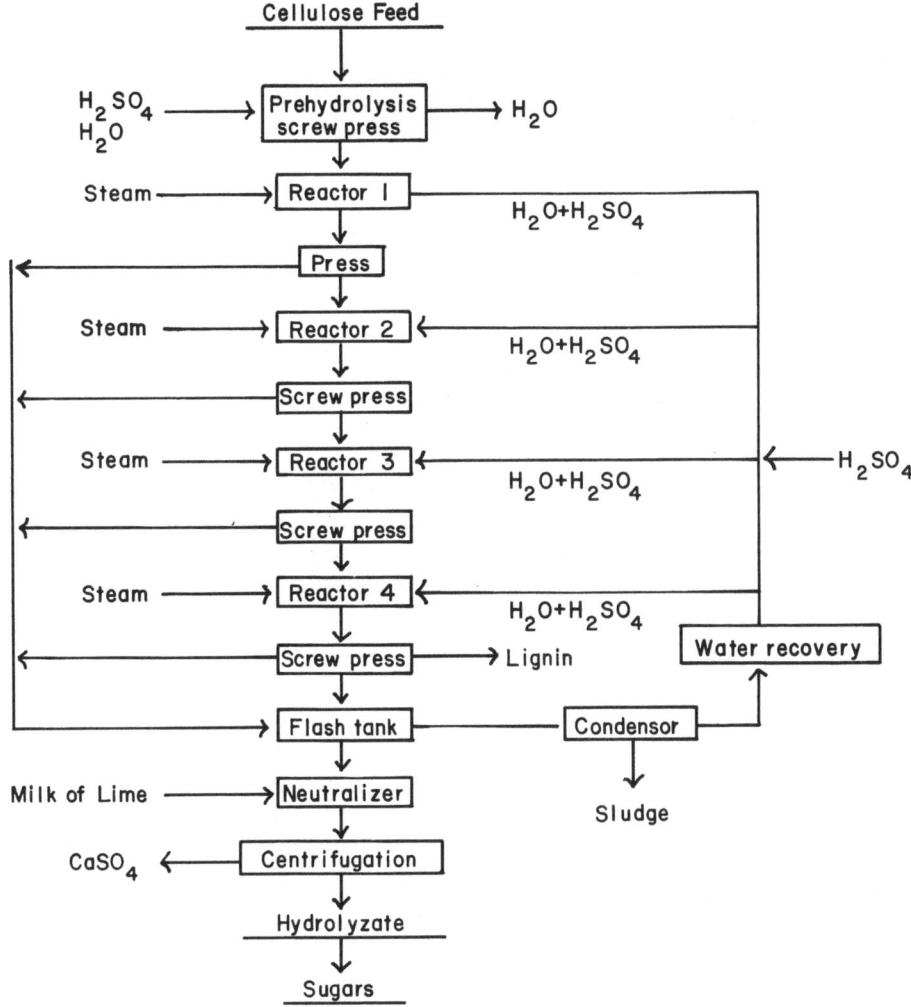

Fig. 5.21. IONICS continuous saccharification process [55]

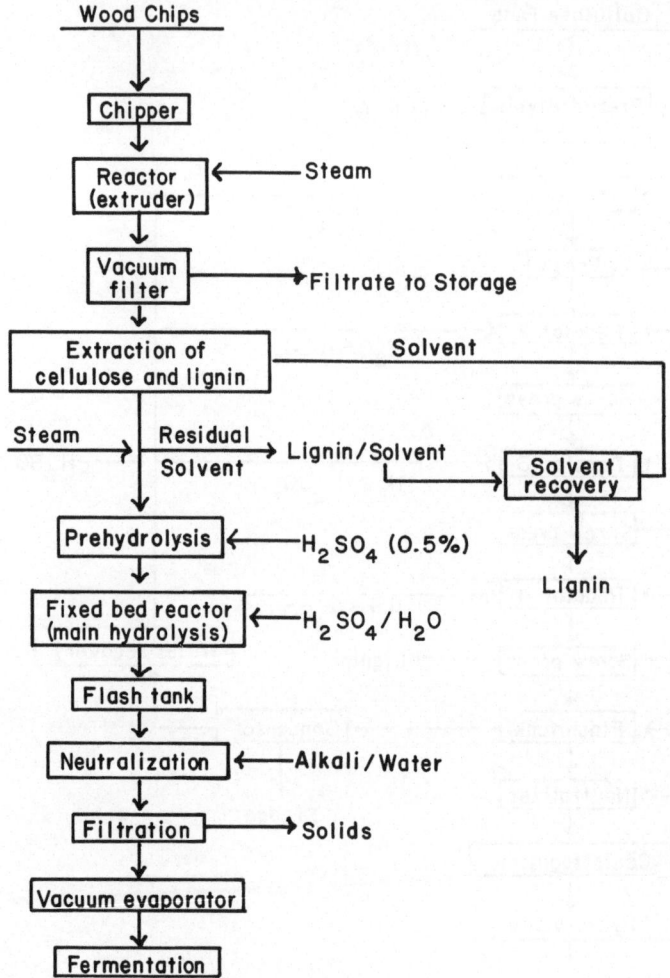

Fig. 5.22. GIT cellulose hydrolysis process [7]

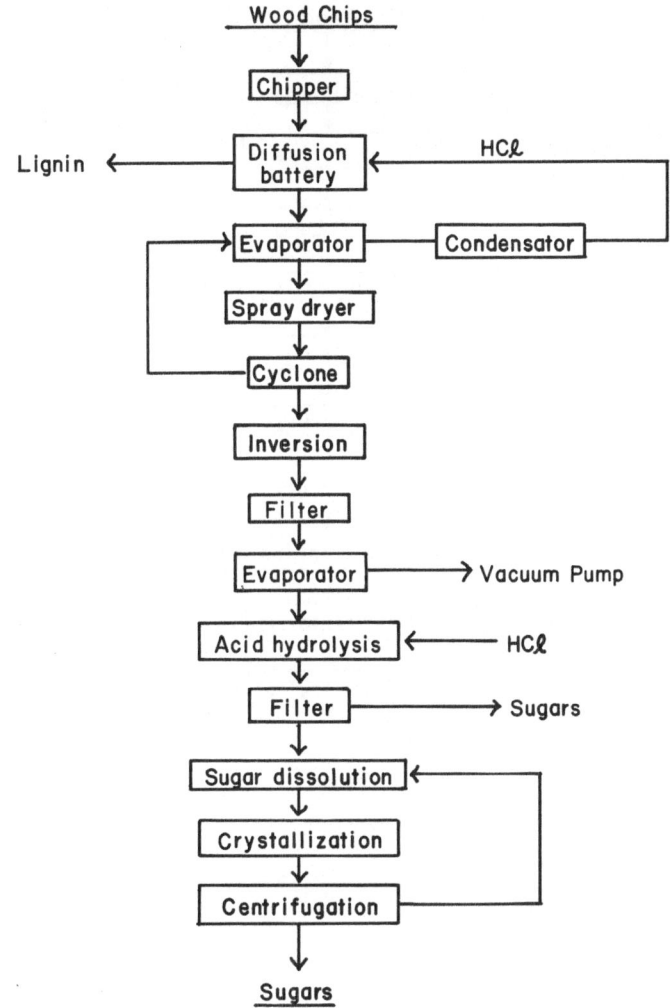

Fig. 5.23. Rheinau wood hydrolysis process [79]

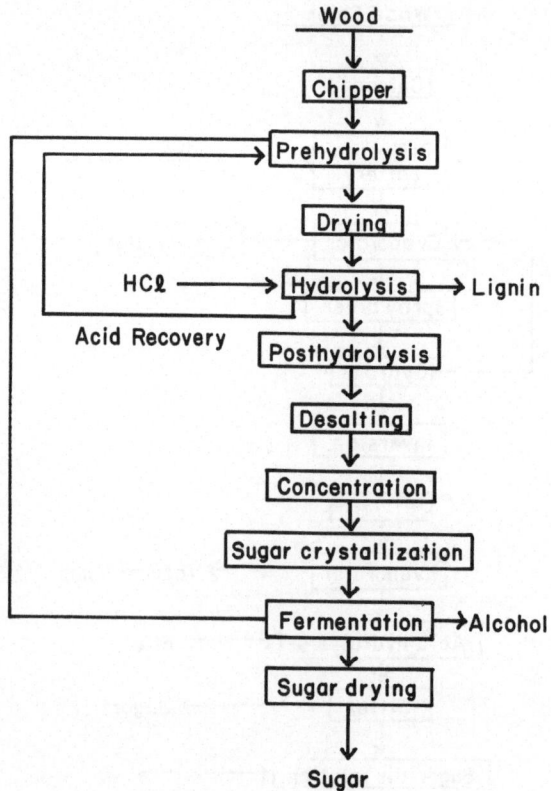

Fig. 5.24. Modified Rheinau process [79]

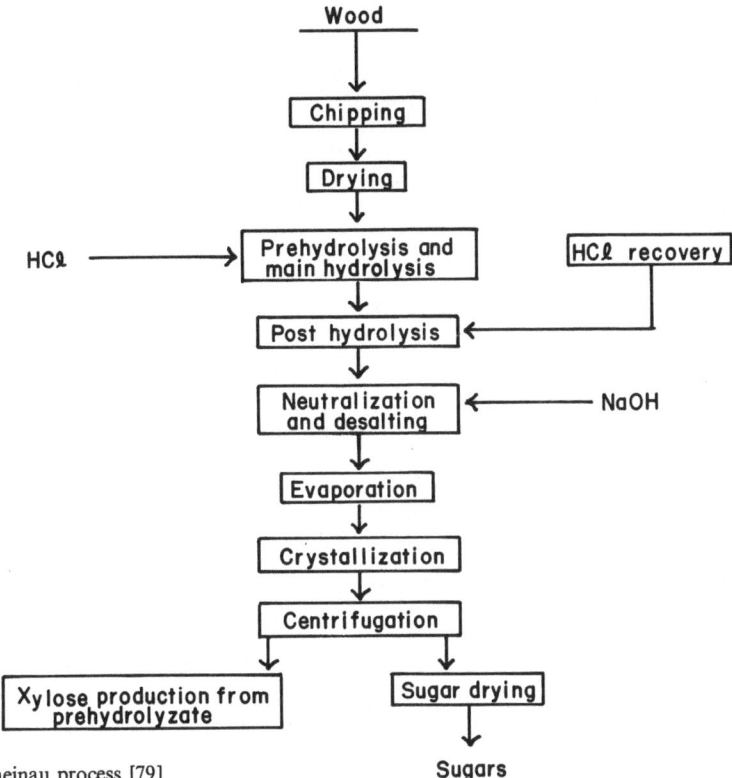

Fig. 5.25. Udic-Rheinau process [79]

Nomenclature

$(E)_0$	= initial concentration of enzyme [Eq. (5.11)]
n_b	= number of batches per day [Eq. (5.6)]
(P)	= concentration of reducing sugars [Eqs. (5.1), (5.3), (5.5), (5.7), (5.9), (5.10), and (5.11)]
(P_1)	= concentration of glucose [Eqs. (5.2), (5.4), (5.8), and (5.9)]
$(P)^*$	= desired concentration of reducing sugar for fermentation [Eq. (5.11)]
R_p	= rate of production of reducing sugars [Eqs. (5.1), (5.3), (5.7), and (5.9)]
R_{P_1}	= rate of production of glucose [Eqs. (5.2), (5.4), (5.8), and (5.9)]
$(S)_0$	= initial concentration of substrate [Eq. (5.11)]
(S)	= concentration of substrate [Eq. (5.11)]
$\$_c$	= cost of cellulose [Eq. (5.11)]
$\$_E$	= cost of enzyme [Eq. (5.11)]
$\$_R$	= cost of reactor [Eq. (5.11)]
$\$_{conc}$	= cost of multieffect evaporation [Eq. (5.11)]
t_b	= batch time in hours [Eq. (5.6)]
TPD	= production capacity for reducing sugars in metric tons per day [Eqs. (5.6) and (5.10)]
V_b	= volume of batch reactor [Eq. (5.5)]
V_{CSTR}	= volume of continuously stirred tank reactor [Eqs. (5.7) and (5.8)]
V_0	= volumetric flow rate of the substrate stream [Eqs. (5.10) and (5.11)]
V_R	= volume of reactor, batch or continuously stirred tank reactor [Eq. (5.11)]

185

References

1. Ackerson M, Ziobro M (1981) Presented at the 3rd Symp Biotechnology in Energy Production and Conservation, Gatlinburg, Tennessee
2. Anonymous (1982) Chem Eng, Feb. 9
3. Anonymous (1982) Chem Eng News, Feb. 1, p 22
4. Bose B et al. (1969) Indian J Technol 7:256
5. Boyle TJ, Tobias MG (1972) TAPPI 55:1247
6. Brenner W et al. (1979) US Environ Prot Agency, Off Res Dev Rep EPA-600-19-79-023b
7. Brooks RE et al. (1980) Contractors' reports, Biomass Refining Newsletter, Summer
8. Casey JP (1960) In: Pulp and Paper, vol 1. Pergamon, New York
9. Centola G (1954) Presented at the 6th Meeting of the FAO Technical Panel on Wood Chemistry Stockholm (1953): FAO of the UN, Rome, Italy, p 77
10. Chalov NV (1959) USSR Patent, 119, 491, April
11. Chalov NV, Blinova NN (1964) Izv Vysskikh Uchebn Zavadenii Lesn Zh 7 (4):132
12. Chalov NV, Blinova NN (1964) ibid. 7 (7):164
13. Chalov NV, Blinova NN (1966) ibid. 9 (3):140
14. Chalov NV, Goryachikh EF (1963) ibid. 6 (1):137
15. Chalov NV, Lappo-Danilevskii VK (1963) ibid. 6 (1):156
16. Chalov NV, Leshchuk AE (1962) ibid. 5 (1):155
17. Clements LD et al. (1982) In: Fuel Grade Ethanol from Cotton Gin Residues, Texas Energy and Natural Resources Advisory Council
18. Collins C (1944) Chem & Met Eng 51:100
19. Converse AO et al. (1973) National Technical Information Series (Springfield, Va.) Report No PB-221-239
20. Dunlap CE et al. (1976) AIChE Symp Ser, no 158 72(158):58
21. Efros IN (1964) Kompleksn Ispolz Drev Petrozavodsk Sb 277
22. Epshtem VV (1975) Gidroliz Lesokhim Prom. 7:1
23. Fagan RD et al. (1971) Environ Sci Technol 5:545
24. Faith WL (1945) Ind Eng Chem 37:9
25. Fong WS et al. (1980) Chem Eng Prog 76 (11):39
26. Gharpuray MM et al. (1983) In: Soltes EJ (ed) Wood and agricultural residues. Research on Use for Feed, Fuels, and Chemicals. Academic Press, New York, p 369
27. Gilbert N et al. (1952) Ind Eng Chem 44:1712
28. GIT (1980) Biomass Refining Newsletter, pp 3, 12, Summer
29. Goto K et al. (1975) J Ferment Technol 53:664
30. Grethlein HE (1978) J Appl Chem Biotechnol 28:296
31. Grethlein HE (1978) Proc 3rd Joint US/USSR Enzyme Engineering Seminar, US, NITS Document PB-283-328-T, Feb., p 441
32. Grethlein HE (1978) Biotechnol Bioeng 20:503
33. Grethlein HE, Converse AO (1979) Presented at the 72nd AIChE Annual Meeting, San Francisco, CA, Nov. 25–29
34. Harris EE, Beglinzer E (1946) Ind Eng Chem 38:890
35. Heidt LJ, Purves CB (1944) J Am Chem Soc 66:1385
36. Humphrey AE (1979) Adv Chem Ser 181
37. Jones JL, Semrau KT (1982) Presented at the 1982 Annual Meeting of the American Institute of Chemical Engineers, Los Angeles, CA, Nov. 14–19
38. Kobayashi T (1954) Japanese Patent 206, 423
39. Kobayashi T (1955/1956) Japanese Patent 219, 160; 231, 289
40. Kobayashi T (1960) FAO technical panel on Wood Chemistry, Tokyo
41. Kobayashi T et al. (1962) J Ferment Technol 40:406
42. Kobayashi T et al. (1964) ibid. 43:852
43. Kobayashi T et al. (1962) Agri Biol Chem 26:9378
44. Korolkov LI (1962) Khim pereabotka Drev Nauchn-Tekhn Sb 11:31
45. Kozlov AI et al. (1965) Sb Tr Vses Nauchn-Issled Inst Gidrolizn i Sufitno Sprit Prom 14:42
46. Kusama J (1979) Chem Econ Engin Rev 11 (16):32

47. Ladisch MR (1971) Proc Biochem 14:21
48. Lee YH et al. (1980) In: Fiechter A (ed) Adv Biochem Eng, vol 17. Springer, Berlin Heidelberg New York, p 131
49. Lee YH et al. (1981) In: Moo-Young M (ed) Advances in biotechnology. Pergamon, Toronto
50. Lee YH et al. (1982) In: Scott CD (ed) Biotechnol Bioeng Symp, no 12. Interscience, New York, p 121
51. Lee YH (1981) In: Enzymatic hydrolysis of insoluble cellulose – a kinetic study. Ph. D. Dissertation, Kansas State University, Manhattan
52. Locke EG, Garnum E (1961) Forest Prod J 11:380
53. Lyons TP (1981) Gasohol USA 3 (2):14
54. Malyshev DA (1965) Gidrolizn i Lesokhim Prom 18 (5):24
55. Meller FH (1969) Conversion of organic waste into yeast. Ionics Inc
56. Odinov PN et al. (1952) USSR Patent 1, 052, 782
57. O'Neil DJ et al. (1979) U.S. Department of Energy Contract, Jan.
58. O'Neil DJ et al. (1979) In: U.S. Department of Energy Contract, Technical Progress Report, no 2, April
59. O'Neil DJ et al. (1979) In: U.S. Department of Energy Contract, no 3, August
60. Oshima M (1965) Wood chemistry, process engineering aspects. Noyes Develop. Corp., New York
61. Peters MS, Timmerhaus KD (1980) Plant design and economics for chemical engineers. McGraw-Hill, New York
62. Riehm Th (1960) FAO Technical Panel on Wood Chemistry, Tokyo
63. Roberts RS et al. (1980) In: Scott CD (ed) Biotechnol Bioeng Symp, no 10. Interscience, New York, p 125
64. Roberts RS et al. (1979) Proc 2nd Symp Biotechnol Energy Production and Conversion, Gatlinburg, TN, Oct. 3–5
65. Rowland SP et al. (1973) J Text Res 43:351
66. Saeman JF (1945) Ind Eng Chem 37:43
67. Sakai Y (1966) Bull Chem Soc (Japan) 39:1036
68. Schoenemann K (1954) In: FAO Technical Panel on Wood Chemistry, Stockholm (1953); FAO Report 54/2/767
69. Sharkov VI, Levanova VP (1960) Gidrolizn i Lesokhim Prom 13 (1):5
70. Sharkov VI et al. (1969) ibid. 21:3
71. Sitton OC et al. (1979) Chem Eng Progress 75 (12):52
72. Sobolevskii CA (1959) Trudy Inst Lesokhoz Problems Akad Nauk Latv SSR 12:113
73. Song SK, Lee YY (1981) Presented at the 2nd World Congress of Chemical Engineering, Montreal, Canada
74. Tanaka M et al. (1980) J Ferment Technol 58:517
75. Taubin BM (1960) Trudy Inst Lesokhoz Problem Akad Nauk Larv SSR 19:159
76. Thompson DR (1977) Thesis, Thayer School of Engineering, Dartmouth College, Hanover, NH
77. Thompson DR, Grethlein HE (1979) Ind Eng Chem Product Res Dev 18:166
78. Vernet G (1923) Chimie Industrie Spec, no 654, May
79. Wenzl HFJ (1970) Chemical technology of wood. Academic Press, New York London
80. Wilke CR, Mitra G (1975) In: Wilke CR (ed) Biotechnol Bioeng Symp, no 5. Interscience, New York, p 253
81. Willstatter R (1913) German Patent 273, 800
82. Wright JD, d'Agincourt CG (1984) In: Evaluation of Sulfuric Acid Hydrolysis Processes for Alcohol Fuel Production, SERI/TR-231-2074, Golden, Colorado

6 Epilogue

Numerous and complex factors need be taken into account in economically assessing processes for hydrolytic utilization of cellulose. Cellulose hydrolysis must compete with hydrolysis of starch from grain crops. While lignocellulosic residues are abundant, their distributions are diffused; the availability of commercially viable collection systems for them is limited. Of all crops, corn is a primary source of starch because its supply is ample, its cost is relatively low, and commerically viable systems are on hand for storing and transporting it over long distances. Cellulosic biomass, approximately at US $ 30 per dry ton currently, is far cheaper than corn approximately at US $ 100 per dry ton. Nevertheless, it is very difficult for most naturally occurring organisms to hydrolyze cellulose due to the inaccessible nature of cellulose crystallites. Consequently, a tradeoff exists between the low raw material cost and high investment cost for cellulose hydrolysis. To enhance hydrolyzability, a cellulosic biomass need be subjected to a variety of pretreatments; however, this increases the cost of processing.

Only a small fraction of crop residues, such as sugar cane bagasse, cotton gin trash, and rice hulls, are presently collected annually. Vast amounts of corn stalks and cereal straw are available and could be collected if demand warrants it. Other agricultural residues are scattered and can not be collected economically or they must be retained on the land to prevent soil erosion. Urban solid waste may constitute a substantial supply of raw material, but the heterogeneity of this material leads to safety and processing problems in downstream processing. The "grassland" cellulose resource, mainly animal manure, is too diffuse, except that found in a few large feedlots. Cereal straw and corn stover – the major agricultural residues – have no infrastructure for collection and handling.

At present, cellulose hydrolysis does not appear to be economically competitive with starch hydrolysis as a source of sugar. Currently, it may be economical only in regions where abundant biomass exists in dense and concentrated form but sources of starch are scarce; the vast forest of Northern Soviet Union is such a region. Processes using concentrated acids to catalyze the hydrolysis of cellulose have been commercially unsuccessful because of the need to recover and recycle the acid. Dilute acid processes reduce acid-associated cost, but the yields from these processes are usually poor. The processes require rigid control of residence time at high temperature; consequently, their energy and investment costs are excessive compared to corn hydrolysis. Enzymatic hydrolysis possesses many desirable characteristics, such as product specificity and less drastic reaction conditions. Unfortunately, the reaction proceeds at an extremely slow rate. To enhance it necessitates pretreatment of the substrate, thereby adding to the processing cost.

Furthermore, the difficult task of enzyme recovery is essential to lower the processing cost.

Currently, the global economy is overly dependent upon oil, a resource that is depleting rapidly. Furthermore, the localized nature of oil resources coupled with political instability in the oil-rich region render the entire global economy unstable. The present oil glut is signalling an entirely misleading picture of world oil reserve. We must realize the forces resulting from diminishing oil reserves will overcome those due to decreasing demand, thereby leading to ever increasing oil prices. Thus, we see the need for alternate energy resources; desirably, they should be renewable. Most developing countries expend their precious foreign currency reserve for oil imports. Energy supplementation plans in these countries would not only curtail foreign dependency but also promote domestic employment.

Author Index

195

Subject Index